ip 26.00 6 50

Studies Show

Studies Show

A Popular Guide to Understanding Scientific Studies

John H. Fennick

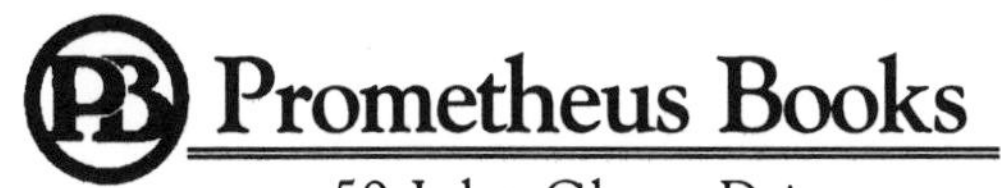

59 John Glenn Drive
Amherst, NewYork 14228-2197

Published 1997 by Prometheus Books

Studies Show: A Popular Guide to Understanding Scientific Studies. Copyright © 1997 by John H. Fennick. All rights reserved. No part of this publication may be reproduced, stored in a retrieval system, or transmitted in any form or by any means, electronic, mechanical, photocopying, recording, or otherwise, without prior written permission of the publisher, except in the case of brief quotations embodied in critical articles and reviews. Inquiries should be addressed to Prometheus Books, 59 John Glenn Drive, Amherst, New York 14228–2197, 716–691–0133. FAX: 716–691–0137.

01 00 99 98 97 5 4 3 2 1

Library of Congress Cataloging-in-Publication Data

Fennick, John.
Studies show— : a popular guide to understanding scientific studies / John H. Fennick.
p. cm.
Includes bibliographical references.
ISBN 1–57392–136–X (alk. paper)
1. Science—Statistical methods—Miscellanea. 2. Statistics—Miscellanea.
I. Title.
Q175.F39 1997
001.4'22—dc21 96–3486
CIP

Printed in the United States of America on acid-free paper

Contents

Foreword

Whenever I teach an introductory statistics course I ask the students to tell me what they think statistics is. Invariably, among the answers is the statement "With statistics you can prove anything you want to." I then go to great lengths to explain to them that with statistics one cannot *prove* anything, but that statistics, if used properly, can indeed demonstrate certain degrees of relationship among factors or variables and that these relationships may sometimes even be causal. The key here is "if used properly," for no discipline has been so misused as statistics.

This is indeed the theme of this book. With a tongue-in-cheek approach the author tries to explain why not every study reported in the media should be taken at face value and that a certain amount of skepticism is appropriate, and, indeed, necessary. In a sometimes amusing way, the author blames much of the confusion about the results of such reports on statistical lingo and faulty data collection. There is, of course, much truth to that, but as a professional statistician I am somewhat concerned that this reflects negatively on my profession, and unjustifiably so, I hasten to add. The reader should be warned that when the author refers to "statisticians" he really means "pseudostatisticians," i.e., "researchers" who have just enough knowledge of statistics to be dangerous. (Statistics seems to be the one field in which someone who has taken only a single course can call himself or herself a statistician, and often does.) So it is only right to warn against unsound practices and to show what is wrong. The author succeeds quite nicely in this effort, but in doing so he, too, sometimes commits a "sin," namely, being too simplistic, and making it sound as if statistics is the easiest thing in the world. Let me assure the reader that this is not true. That does not mean, of course, that one should not try to make the subject more easily understandable for the layperson. And to the extent that this can be fun and entertaining, this book is a step in the right direction. It should open some eyes and make for interesting reading.

Klaus Hinkelmann
Professor of Statistics
Virginia Tech

Preface

For more than forty years we have been bombarded with "results of studies." Statistical study reports convey the latest dietary, medical, environmental, or other discovery for instant implementation into our daily routines.

By now of course, we know that the "latest" study will often contradict those that have gone before. Depending on an individual's makeup, one may find that to be amusing or highly irritating. In either case however, one is never quite certain if it's okay to sugar the coffee, or even to *drink* the coffee.

Is this consistent inconsistency just the inherent nature of statistics, a consequence of the pursuit of grants and the "Publish or Perish Syndrome,"* or something else entirely? While publication of dubious studies is spurred by the struggle for funding and the lure of catchy headlines, I believe it occurs because most practitioners simply don't understand the statistical tools they use, nor have they received adequate training in their use. Often researchers have no appreciation for the vagaries of the particular method they have chosen to examine their data. They have no idea of how minor changes in that data can produce totally different results. Nor do they question the validity of whatever answers the computer pumps

---

*Researchers, worried about grant renewals, are commonly under pressure to publish any result, good, bad, or indifferent. This urgency is encapsulated in the phrase: "Publish or perish."

out. It seems they willingly accept whatever numbers their analyses produce, and even the "experts" accept repetitive contradictions as normal. This fact is obvious even to the layperson: Take the extreme case of demonstrating that lefthanded people die nine months earlier than righthanded ones and then, when challenged, publishing a second result showing that the first was in error by eight years. It's rather sad that the problems with all studies are not that obvious.

The uncertain ground under any statistical study was commonly acknowledged in early research papers published in the 1950s, and often discussed in detail in reports published through the 1970s. Writers, as well as readers, got weary of repeating the same old cautions and warnings, so their appearance dramatically diminished. Now, given the volume of reports and the lack of such warnings, the uncertainties are forgotten and even the practitioners come to believe the results. This conditioning of the analysts, as well as the public, is generating contempt for statistical studies in particular, and science in general. The situation is framed by a colleague's recent remark: "*The New England Journal of Medicine*? Oh, the *National Enquirer* of the medical world."

As I see it, the problem begins with the teaching of statistics. In general, statistics textbooks lack any reasonable explanations or understandable justifications for the tools they propose to teach; they commonly dwell on mathematical manipulations and discuss the subtleties of things like "independence" or "stationarity" while ignoring practical limitations. All of this is done in a language that few *post-graduates* understand. There appears to be an unwritten law applied to the authors forbidding comprehensible prose. The future researchers in botany, medicine, psychology, or any other discipline nod their heads in confused unison, then go on to make their own contributions to the pile of published heresy.

Everyone has heard about lying with statistics. That's easy to do and I will show some minor examples. It is quite another thing, however, when the analysts put their faith in methodologies and practices that yield invalid results, simply because the error estimates are small. No one has ever told them that the error estimates are themselves subject to gross inaccuracies or that what error they choose to accept is subjective and even capricious. They fall into their own version of the popular wisdom, "if the computer says so, it must be right."

This book is not *on* statistics but *about* them. It is a book, in plain English, that incorporates mathematics no more complicated than arithmetic, yet it touches almost every area relevant to conflicting scientific reports. There is no attempt to explain *how to do* statistics, but I hope to instill a healthy *understanding* of it.

By describing the shortcomings, pitfalls, and traps with glaring examples and a touch of humor, it is hoped that the lessons will stick. If digested early on, this book should also promote a willingness to learn the "how" by discovering the "why" beforehand. If it is read before attempting a text on statistics, readers should find it refreshing to know the direction, the whys, and the wherefores from the start. In addition, I hope that the average layperson will cultivate a healthy skepticism when confronted with the latest "study."

In this book I introduce you to a fair amount of detail about statistical studies. When you see for yourself some major pitfalls of statistical analysis and how easy it is trip on them, hopefully you can make wiser judgments about the advice that pours from the daily health columns.

Before delving into all of that however, I wish to point to two broad problems of a fundamental nature that are not addressed in the text, but that seriously impair our ability to make wise judgments. First is the kind and quality of information in the original study reports, second is a collection of popular misconceptions.

First, What is in the study reports:

Research reports are written by and for knowledgeable peers. There is no need for lengthy explanations and results and conclusions are brief and candid. It is important to note that there is a difference between results and conclusions, both in content and as to who pays attention to them. That last remark is crucial to what I mean by the kind of information the reports contain. Results are technical, hard to understand, and are read by the experts; conclusions are subjective, easily misleading, and read by laymen. Consider the following excerpts from a typical study:

> Results. After allowing for age . . . risk of myocardial infarction [heart attack] increased as the degree of vertex baldness increased ($P < .01$); for severe vertex baldness the relative risk was 3.4 (95% confidence interval, 1.7 to 7.0).

> Conclusion. These data support the hypothesis that male pattern baldness involving the vertex scalp is associated with coronary heart disease in men under the age of 55 years.*

As you will learn in this book, those results say no more than that there may be something of interest here but we generated no real evidence. Other researchers will read that and more than likely skip to the next article. Reporters or laymen scanning the journals will read only the conclusion, because that is what they can understand. In this particular instance, on the very day this report appeared in the *Journal of the American Medical Association,* an AP news release bore the headline "Bald Head, Bad Heart? Could Be" and went on to say that men, severely bald at the top, were three times more likely to suffer heart attacks. If you took the time to read the entire article, you would learn that the study was funded by the manufacturer of a baldness treatment product, that eight prior studies had been inconclusive and that the team from the Boston University School of Medicine that conducted the study knew of no reason for the *apparent* link between heart disease and baldness.

The researchers who wrote and read the report may have been shocked at the reaction in the press because they "know" that nothing like that headline was said or intended. (More likely, if they even read the newspaper article, they shrugged it off as "pop" journalism.) When they read reports, researchers look for and find only the factual content. A few key words indicate what is interesting and what is meaningless. There is no need to *read the words,* each researcher fills in his own as the facts are gleaned. As a consequence, what researchers see in published journals may be very different from what is reported in the press. What reporters may headline as the latest breakthrough carries no more weight for the researcher than the first quarter score of a football game. It is a simple status report from one of the laboratories. If a reporter wants to call it a breakthrough, that's his problem. There is no point in trying to correct the press. Gross statements in the media, though they may be irritating, are left unchallenged. It is not as though scientists are cavalier as to how their efforts are interpreted, but when it comes to publicizing their work, those who avoid the spotlight or clarify their results are rare.

---

*Samuel M. Lesko, Lynn Rosenberg and Samuel Shapiro, "A Case-Control Study of Baldness in Relation to Myocardial Infarction in Men," *Journal of the American Medical Association* 269, no. 8 (February 24, 1993): 998–1003.

This is a vulnerable scene, with a vested public interest, that has been exploited by media opportunists for decades. Not surprisingly, there is a just and rising distrust of the scientific community. In keeping with this spirit of rising distrust, within these pages you will find facts and comments that seem disparaging but simply reflect my reactions to many journal articles, as well as columnists' reports. These are intended as good-natured jabs. If any researchers are offended, I trust they are not also among those who complain about reductions in research support. If the public is financing work, the public must expect to be properly informed. In the sensitive and highly technical arena of medical research, the bulk of that task falls on the worker; few are capable of translating his words.

To aggravate the problems of misinterpretation, there are practical aspects of report writing that encourage drawing conclusions of the type cited above. Scientists worry as much as anyone about income stability, whether it comes from a pharmaceutical company paycheck or a federal grant. When reports are written, there is no great desire to publish negative results. That's not the way to advance or to ensure the grant continuance. In no way does this mean that reports need be falsified; the above conclusion is a true statement about baldness and heart attacks. However, it would have been equally true to say, in spite of these indecisive results, we cannot argue that the data, analyzed in this manner, do not support the hypothesis that severe baldness and heart disease are related.

Simple changes in the written presentation can divert the reader's attention from negative aspects. It is then no surprise that those aspects never appear in later condensations. This is human nature at work, but it can cost the end reader vital information.

The same principle applies on a larger scale to reports coming from the universities or federal or private "independent" agencies. Practically everyone has a pet peeve, an ax to grind, or cause to champion. Pure objectivity is hard to find. Even if personal, political, and financial biases are controlled, any report has a limited amount of available space, some detail must be omitted, and someone must decide what is least important. That definition is forever subject to change. What is legitimately dismissed as inconsequential today may be later discovered as the key to a puzzle. Once again, critical information is lost.

And so we have the sad but realistic result that the very reports relied upon for the truth about health and medicine are intrinsically stacked against us because of our lack of knowledge.

Now for the misconceptions:

Those comments on the natural biases in human nature are probably not news to you and you may have been suspicious of past reports because of similar ideas. There are other traps for the unwary, however.

First is the cloak of respectability. Reports in prestigious publications or from renowned organizations tend to be treated as gospel. Scientific journals often command respect because they are billed as publishers of peer-reviewed articles. The understanding is that the journal articles are more trustworthy because of the peer review. Many journals subject all submitted articles to the peer review process, and state that fact in their "information for authors" section. Others leave the question open. The *New England Journal of Medicine,* for example, often touted as one of the best, warns prospective authors that their articles may be subjected to peer review. In their words "manuscripts are . . . usually sent to outside reviewers." However, published articles are not identified as being reviewed or not. When journals such as these are referenced, you cannot trust that any truly knowledgeable person, other than the author, has inspected the work for quality or integrity. You never know when the cloak of respectability has a hole in it.

Also among the popular misconceptions is the lore that studies employing thousands of participants are just naturally better than those using only a few. In some cases there is an element of truth to this, but the headlines stating that 20,000 Harvard graduates or 100,000 nurses participated in a study of heart disease is a waste of newspaper print. This question is examined in detail in lesson 6. Briefly however, how many healthy people participate is of little import. What is important is the number of cases of heart disease. In general one does not need 20,000 or 100,000 subjects to locate a few hundred heart attacks, and a few hundred cases is usually adequate. In the less common studies of rare diseases such as childhood leukemia, thousands of subjects are needed simply to locate even a small number of cases. That is a different matter altogether.

There is another problem with thousands of subjects and studies that continue for decades: the data collection process. If a study is to be at all credible, the data from which it draws conclusions must be impeccable. But, consider questions such as alcohol intake or use of contraceptives by up to 100,000 nurses over twenty years. The subjects are not given frequent physical exams or even brought to some screening and interview facility for in-depth questioning on these matters. That would be unbearably expensive and time consuming. Use of the items is based on questionnaires, in most cases sent out at intervals of years. The subjects are asked to recall how much they drank on a daily or weekly basis over the past several years, or how religiously they used oral contraceptives. Often several paragraphs in the reports of these studies, and sometimes entire reports, are devoted to arguments that the subjects' responses to the questionnaires "really are reliable." In any case, these questionnaires, answered through human recall, then become the basis of scary reports that this amount of alcohol is good or bad for your heart. Some even have the gall to say *how* good or bad by citing risk percentages. One brighter note on this particular problem: For some reason, columnists seem to like to tell you when the study was based on questionnaires, so the warnings may come with the report in this instance. You don't have to search for the fact that the underlying information is suspect.

So, adding to the biases developed from misreading reports, we have our built-in misconceptions about peer reviews and thousands of subjects that set us up to be misled.

Before closing these preliminary remarks it is appropriate to address another topic of concern to many: the number of studies conducted on the same subject. Even if we dismiss the many bad ones as not counting, it is fair to ask how many studies are enough? For example, one gets the impression that the number of smoking-cancer studies belongs in an astronomer's handbook, it is so large.

There are two answers to this question. In the case where successive studies agree, good statisticians, reviewing the evidence with experts on the condition under scrutiny, can easily decide when further studies will not add to the current knowledge. On the other hand, when many studies disagree or come up as indecisive, it is time to quit. As we will see in several places in the text, this is a convincing sign that there are fundamental problems with the studies already conducted, and until they are understood, further studies of a similar type are a waste of resources.

In either case, there is never real justification for racking up hundreds of studies on the same subject. Time and resources would be much better spent elsewhere. The National Academy of Sciences recently "took the bull by the horns" (as best they could) on this issue, with the cancer-power line studies. They reviewed the hundreds that have been done, recognized that they were going nowhere, and declared that no risk could be found. They did not call for a moratorium, and a few on the panel even objected to the recommendation published. However, they at least responded to the need to relieve public anxiety. This last "study" was the best one in this area.

Finally, I wish to be specific about the types of studies of interest in this book. We are concerned with serious medical studies, not the longevity of left handers, or whether a Mediterranean diet will make you live longer. I will however, refer to several of this latter type to illustrate points.

The Scheme of Things

My plan is to explain statistics at the primary level. A real effort has been made to keep everything simple yet comprehensive. I have assiduously avoided the deep water while trying to cover much of the shoreline. To this end, I have taken considerable liberties with the nomenclature and, by the end of the book you'll find that I prefer my terms. Having known the occasional statistician with a real sense of humor, I'd wager that they might agree.

My first such liberty is dividing my book into lessons rather than chapters, to remind you this material comprises the practical lessons omitted in the standard texts. These brief lessons comprise a mix of anecdotes, case histories, and tutorial passages explaining statistical measures. I hope you will find the anecdotes entertaining and instructional; the case histories sad or amusing, but educational; and the tutorials stimulating. We proceed from the very light, tongue-in-cheek approach of lesson 1—a kind of "Statistics According to Me"—to the serious critique of some of the statisticians' most cherished tools in lessons 6 and 7.

The book is intended for two very specific audiences. The first is anyone who might appreciate an accessible understanding of why "scientific studies" come across as so unscientific. The second is the student of statistics. To those in the first group, this book should answer your questions about statistics without the need for an advanced degree. For the second group, the following is just for you.

To the Student

Most texts, beginner's or otherwise, concentrate on telling you how to do something, but seldom make it clear what it is you are doing.

The true goal of the task is rarely set forth. It is like exhorting the worker to mix fine concrete but never telling him he is building a Taj Mahal. The emphasis here is reversed. I want you to know what statisticians do, how good the tools really are, and how to evaluate results. If this book excites you enough, you can read any number of standard texts for the details on how to do statistics.

And, once more:

For the General Reader

Please understand that this is not a textbook. It is an informational dialog between you and me; even if you can't reply directly. I am more at ease in writing this sort of material if I feel that I am talking with you rather than at you. I also believe that you will enjoy the material more with that approach.

In reading *Studies Show,* I hope you will benefit by "learning," in a most informal manner, why the health gurus cannot decide if the cholesterol in eggs is or is not bad, or whether lobsters even have any real cholesterol. "Learning" is in quotes because you will "learn" very little about statistics, in the classroom sense. However, you should come away from the book with a better understanding of the subject than many so-called statistical researchers.

I have attempted to keep the first five lessons as light as possible, providing examples that try to explain some mathematical definitions and practices yet avoid the math behind them. This was not always possible, and you will have to trust that I am not lying to you if I just say, in essence, that "this is the way it is." Of course, you could seek your own expert for confirmation. In the closing lessons, where the real shortcomings of statistical studies are laid out, the going gets a little tougher. It borders on being unduly technical and you may have to work a little to comprehend. If you are willing to make the effort, you will be as shocked as I was when I discovered that this is the way some "researchers" actually carry on, using tools they don't comprehend.

Introduction

Did you know that living near the equator is dangerous to your health? It's true. My studies show that on average, people live about a half year longer for each 100 miles of resident distance from the equator. Here in the upper hemisphere, this means that people who live 1,000 miles north from wherever you do have a life expectancy five years greater than yours. That's equal to what we supposedly gained from the past thirty years of medical progress. Does this mean that if you move north you will live longer? What do you think?

"Stupid," you say. "Of course not." Yet, the evidence supporting those facts is as good or better than that supporting many statistical "truths" that we sustain with billions of dollars of grant money each year. For example, it is considerably better than the evidence that low-fat diets and exercise have reduced deadly heart attacks. So, by this measure, my study should be in the headlines.

Statistical studies like mine have been making headlines for years. That's okay I guess, but their authoritarian omens of doom or glory berate our habits and shuffle our lifestyles. In mortal fear, we lie defenseless in a wealth of statistical ignorance.

Here is your chance to change all that. With all the noise about scientific studies and their counter-counterclaims, it's high time someone explained them, in a way that people can understand. In the following pages you will discover what statistical researchers do, why their results conflict, and you may actually learn something about statistics. As you learn, you may also find that the nasty thoughts you had about all those studies are not completely unfounded.

The intent here is to teach you enough about statistics to let you see through the reports and understand what has been done, rather than just what is said.

Lesson 1

The Rules

For openers, let's get a brief but broad view of the entire field of statistical studies. We need an overview of what's being done and, more important at this point, some of the rationale for why it's being done. To this end, we consider five rules establishing the master framework for all studies, and provide several examples of how each is used. These are my rules but they are nonetheless empirical. I have created them as the only possible explanation for the apparent chaos driving the flood of reports we hear about each year. They are designed to ease you into the world of scientific studies with a familiar and comfortable feeling. To assist in this, I shall attempt to relate each of them to recent or well-known headlines; and I promise, in this first lesson, to be only very slightly technical, not even enough to get your feet wet, just enough to moisten your toes.

Rule 1: Extol or ignore the obvious (whimsically).

This is the "Prince of Rules" for all statistical studies; it provides the user with the gall to publish anything. Pointing to carefully chosen, well-known facts, relevant or otherwise, lends credence to the study results, and outlandish conclusions become palatable. On the other hand, one should never become flustered by obvious facts that belie the claims. Researcher beware: This rule is a two-edged sword. Here are examples of how it works:

Referring to the introduction, regardless of how the result of my study of the equator struck you, consider some corroborating facts (extolling the obvious). Traveling north to south across Europe into Africa, in spite of what you may have heard about the Mediterranean diet, we find that Swedes outlive all central Europeans, who in turn outlive Egyptians. On our side of the Atlantic, Canadians and Americans live longer than Mexicans and, south of the equator, Australians enjoy this earth much longer than the inhabitants of the many nations between them and that central divisor. Even within the United States, the facts are that the death rate in Alaska is about one-half of that in the southern states and it peaks in Florida. Now do you think there may be something to my result?

Including such "obvious" tidbits in any report lends circumstantial proof to the thesis. Little goodies such as these, with which people may already be familiar, beef up the article even though they are coincidental or totally irrelevant. However, if not chosen carefully, a critic or antagonist might swing the other edge of the sword. For example, it could be pointed out that the Japanese live about three years longer than the British, yet London is 1,000 miles north of Tokyo. As for Alaska and Florida, death rate has nothing to do with longevity; Alaska has a relatively young population, while it is common knowledge that many older people move to Florida. The longevity in Alaska is actually about two years less than in Florida. This last stunt, substituting a related statistic for the real one—death rate for longevity here—is a common tactic to trip up the unwary. The truth in this case is that my result is sheer coincidence due to an obvious circumstance that I will (officially) choose to ignore. It so happens that lesser developed countries tend to be closer to the equator, and lesser developed generally means poorer health. The proper conclusion should be that *my study is worthless.* However, wanting very much to publish it, *and* being aware of those contradictory facts, I simply add: "More work will be required before the significance of this result can be evaluated." Sound familiar? Furthermore, if I publish, I might provide Alaskans with another benefit—their real estate could skyrocket.

My equatorial study is statistically sound (I'll provide the data to interested parties). The result is a simple example of lying with statistics. Nevertheless, if I were to formally publish my data, it would be picked up by the media and interpreted as "proof" of a palpably ludicrous "fact." In this case, the fallacies are easily recognized, even those in the corroborating facts. But things aren't always that simple, especially for the poor layperson. Having this sort of thing in mind, an officer at the National Research Council (NRC), working with a group to find better ways of presenting study results to the public, was quoted as saying: "A person should consider all available information before evaluating any published research." This is as helpful as advising that the plans of all terrorist groups should be examined before boarding an airplane. For example, take the flap over power lines and cancer. Picture yourself running around collecting the facts, and then "considering" how many kilovolts and microteslas you prefer? How many what? Don't fret, if you wait until fall, I think they come in more vivid colors. This bit of wisdom from the NRC is simply not practical; I urge you to read on.

Many reported studies, including my own above, begin with the finding of a correlation between two things, e.g., longevity and geographical location. How easy it is to then get carried away in the pursuit of new knowledge and ignore the obvious, such as more developed versus less developed countries. Researchers dabbling in statistics are notorious for this. We will encounter many examples in the following pages. If you are a budding student in a field of research that uses these methods, start to break this habit now.

A good example of how this comes about is a report dating back to 1967.[1] At that point in time, many researchers were personally convinced that high-fat diets were quite bad and that Americans should be severely chastised for liking them. The name of the game was "prove that point." One researcher had been examining the mortality of Japanese Americans (that is, Americans of Japanese ancestry and Japanese immigrants who had been in the United States for more than ten years) and comparing their death rates to white Americans and to Japanese in Japan. His report of that year includes numerous tables and charts demonstrating that when Japanese move to this country their incidence of fatal heart disease jumps dramatically compared to that in Japan. The numbers are really scary.

Now, the researcher left it right there, simply pointing to that terrible jump in fatal heart attacks caused by a move to this country and, supposedly, the high-fat diet. This paper has been referenced over and over again as proof positive that high-fat diets kill people, even the normally resistant Japanese. What was not mentioned in the original paper, though the numbers show it, is that even though the Japanese in this country did have many more fatal heart attacks than their cousins back home, they also lived a lot longer than their kinfolk. Yes, in 1967, Americans lived longer than Japanese. Furthermore, the Japanese Americans were living even longer than the white Americans whose fatty diet they had adopted. Down through the years, these study-mongers continued to ignore one simple fact. They let us believe that the bad American diet was killing off the Japanese just as fast as the Americans when, in fact, if their favorite tool, age-adjusted death rate, was used, *the Japanese were much better off.*

The curse and controversy of the fatty diet dates back to the 1950s.[2] Some researchers were struck by a low incidence of heart disease in arid areas of the Middle East. Painstaking studies revealed that people there ate little fatty meat. (Cattle don't flourish in a desert.) Now, of the hundreds of differences between people and lifestyles of the Middle East and those of the United States, why pick on steak and hamburgers? I haven't the faintest idea, do you?

About twenty years after that brilliant work in the Middle East, researchers were still beating the drums for low-fat diets and for some serious tests to prove their claims (there still aren't any). Then, someone finally noticed something obvious in this case, the contradiction presented by the Eskimo: whale and walrus fat diets, but not much heart trouble. A flurry of activity brought fish oil into prominence—for a short time. Of course you know, the fish oil diversion was only one of the continuing rule changes in the fat-cholesterol game. If you haven't read today's health column, you're probably out of date. And so it goes: Study leads to recommendation leads to study leads to contradiction leads to new study.

Then again, maybe we really shouldn't complain. After all, medical research, full of incomplete and conflicting results, is no different than any other research. If it ever were complete, then the research would be over. However, good research papers, while explaining a little something, invariably open the door to new vistas of ignorance, spurring more research. What has happened here is that now, to feed our quest for the elixir of youth, even the new vistas of ignorance must be put into practice. Which brings us to rule 2.

Rule 2: Promote ignorance as knowledge.

The practice of rule 2 is similar to that of rule 1; they are both variations on telling the "big lie." This one involves dazzling the audience with sleight-of-hand. It's like building a bonfire of nonsense to hide a spark of truth.

The pervasiveness of converting ignorance into medical knowledge was recently driven home to me in a doctor's examination room. On the wall, a pharmaceutical poster extolled the virtues of an anticholesterol medication. To champion the message of the poster, someone had attached an article describing a supporting study. It was like the others, trying to impress you by stating how many thousands of subjects participated in the study and control groups, i.e., those who took the medication and those who did not. How many fewer heart attacks, and by implication saved lives, occurred in the study group, etc., etc. The closing line was unusual, perhaps an effort to lend credibility to a worn-out theme. It said, "Overall mortality in both groups was essentially the same." (There was the spark of truth.) So what killed the people who did not have the heart attacks? The writer, ignoring the obvious, never wondered, so why should anyone else? When I asked the doctor about the apparent conflict—with or without the lower cholesterol, the mortality rate was the same—it was not clear that he even understood the question. His response (the bonfire) was a lecture on risk factors. At the time of this writing, a report claiming that efforts to control fat and cholesterol do not affect longevity was just released.[3] Needless to say, much of the medical profession is up in arms.

The very introduction of the term "risk factor" is a subtle example of putting ignorance into practice. It conjures up an unholy relation between two things: in this case a high-fat diet and the incidence of heart disease. It is particularly useful because it is "self-explanatory," even though it defies definition. (Just ask "how risky?" and see if you understand the answer.) It is glibly applied to a whole host of strong but poorly understood, and mostly meaningless, correlations. "Wait a minute," you say. "What is this correlation stuff?" Good question. Here comes your first real lesson and a taste of the technical.

When we say that two things are correlated, it means that the size or growth of one is closely tied to the size or growth of the other. Later on we will see that we can even tell how closely the two things are tied with a ruler that measures from –1 to +1. At either extreme the things are very closely tied and, at zero, one thing doesn't care that the other exists (most of the time, that is). Here is a trivial example of how correlations work:

Think of a time a long way back and imagine yourself as a Neanderthal witch doctor. You and your people do not understand seasons and your people really hate cold weather. They beseech you to do something about it. Wracking your superior brain, you decide to make some measurements (you're an enlightened witch doctor). For no particular reason, you begin to record the length of time between sunset and sunrise and whether the following day is hot or cold. When you examine your records, you notice that long nights seem to go along with cold days. You have found that these events have a strong correlation. Preempting modern science, you make the obvious pronouncement: Darkness is a risk factor for cold. Your total ignorance of seasons is translated to sacred knowledge under the guise of a risk. Having made this study in prehistoric times, you could have been responsible for the construction of megalithic idols to worship daylight.

"Risk factor" is the accepted terminology today; not many years ago, one would have said that darkness *causes* cold. (The more vehement faddists still do.) "Risk factor" is an indirect acknowledgment that statistics cannot prove anything, they merely suggest a relation. It's important to grasp the subtleties here, "risk factor" instead of "cause." However, if you're the typical American, you tend to believe that "risk factor" equals "cause." I hope this book will clear your head.

In the case of heart disease, by my last count, there were seven major risk factors, including dietary fat, smoking, lack of exercise, and other things people enjoy. Collectively, risk factors "indicate" that, if you were born, then it's likely someday you'll die. The fact that we do die, for whatever reason, supports the proof-of-the-pudding cry, "I told you so!"

You will encounter a lot of discussion about risk factors in this book. I just mentioned that currently there are seven major ones for heart disease. It has been claimed that the workers in the famous on-going Framingham Study* have discovered more than two hundred such risk factors.[4] By the end of lesson 7 you should have a perfect understanding of the facts and fancies behind such remarks and the risks you really take by believing them.

One of the risk factors early associated with heart disease is the "Type A" personality. Herein lies another outstanding example of overlooking the obvious as well as promoting ignorance. The Type A culprit is hard-driving and aggressive, much like the stereotypical entrepreneur of Japan. Now, the longevity of the Japanese and their lack of heart attacks are favorite topics of beef bashers, who choose to extol the fatty risk but ignore the Type A one. The flagrant Japan-

---

*The Framingham Study is discussed in detail in lesson 6.

ese violation of the Type A death sentence has been obvious for decades, yet as recently as 1991, researchers were still beating the Type A-heart attack drum.[5] An article published in that year reported techniques to measure the relative density of Type A people within populations, and results for thirty-odd American cities and six foreign countries (including Japan) were presented. The author proudly displayed his "strong" correlation (+0.5) of heart disease with Type A activity measured in the thirty cities. "Best one yet," he claimed. No mention was made of heart disease in the other countries. However, a check of heart disease death rates in the four of them for which data are available[6] yielded a correlation of –0.4 with the reported Type A densities, almost a complete reversal. Even though the researcher pointed out that the Japanese were consistently faster paced (Type A) than even Americans, apparently following rule 1, he ignored the fact that all those Type A Japanese have a vanishingly small incidence of heart disease. One wonders how the fatty Eskimo diet was ever recognized. I suspect that some researcher wanted to justify his fetish for fish oil. (Don't any of them like rice?)

By now you see that a basic requirement for statistical research is a set of blinders. A good researcher must never be bothered by obvious contradictory observations—my Great Aunt Sophie always did that, and she lived to be ninety-eight! Tracking down reasons for numerous counterexamples could take years, seriously jeopardizing the funding. Not only that, there's the risk factor that says those counterexamples might invalidate the study—even before it is published.

Rule 3: Assign a risk.

Risk factors are fun items and extremely popular. The problem with them, but also a reason for their popularity, is that no one really understands them. As alluded to earlier, they are great bugaboos. The very name smacks of lady luck, ill fortune, the intrigue of gambling. It defies confrontation. It has become a terminator of intelligent discussion. Risk factors are bandied about so much that it behooves us to become familiar with them. Let's explore a few to see what can be learned.

Consider air travel. Worldwide, it incurs about one death for every two billion passenger miles.[7] So what? Is that good, bad, or indifferent? Well, there is a risk factor in air travel and that will explain everything, won't it? Let's ferret one out and see.

One way to state a risk factor is the probability, or the chance, of something happening. In the case of airplanes, does that mean that if you fly two billion miles you'll be dead? (Probably, because you're not likely to live long enough to do that.) No, because that would be a guarantee, not a chance. How about if two billion people fly one mile, then one will be dead? No, that's no good either, it's still a guarantee. Furthermore, that would require two billion single passenger flights of one mile each. Not practical at best. Well, suppose there are ninety coast-to-coast flights, each with two hundred passengers, every day (risk factors can get complicated). Then, in twenty years, there would be about 400 billion airline passenger miles generated (20 years × 365 days × 90 flights × 200 people × 3,000 miles). If, as we would expect, all two hundred passengers were killed when one of these flights crashed, then according to the numbers we could expect exactly one of them to crash during the twenty-year period. (We had to allow twenty years to generate enough passenger miles to kill two hundred people.) Note carefully the word "expect," not "guarantee." The risk factor for such a crash is simply one over the total number of flights, approximately one in 660,000. There, that's a risk factor for planes loaded with two hundred people flying three thousand miles. That doesn't mean that you would be on that plane, however. A more personal way of stating this same risk factor is that every time you take a coast-to-coast flight, there's about one chance in half a million that you won't make it. By the way, that's about sixteen times better than the chance of winning a state lottery. This means, of course, that you can expect to be in sixteen fatal air crashes before you win a lottery. Even if you fly coast to coast every day, your risk for each flight does not change. It's just like the lottery, your chance of winning this week's draw is not affected by whether you played last week. (I know, a lot of you don't believe that.)

Now that you know all about the risk of flying, do you feel better or worse? Come on. You don't have any more feel for that than if I said there were two parts per million of iron in your tap water. So what? Is that better or worse than two parts per billion? By how much? Should you call the Health Department? Obviously you still need help.

Air travel is common; crashes and their tolls are well publicized. How many people have you known who were killed in air crashes? How many people do you know who have made many successful air trips? Thinking about those things gives you a much better feel for the risk of air travel than all the numbers of the example we just went through. Furthermore, whether you enjoy or dread flying, neither the numbers nor your flying buddies are going to change your mind.

If air travel and its tiny risk factor is incomprehensible, how about travel by car? That's something everyone knows about, maybe we can relate to the equivalent numbers for that activity.

As has been well publicized, traffic fatalities have decreased in recent years, making for a rosier picture. Even so, we can estimate from available data that, if the typical car is carrying two passengers, then Americans generate approximately 3,000 billion passenger miles per year. Now, each year about 30,000 passenger car occupants are killed in accidents.[8] That amounts to ten deaths for each billion passenger miles, or twenty deaths for two billion, as opposed to the one death for two billion for the airplanes. So, by this measure, cars are twenty times more deadly than airplanes. Now, would you rather fly or drive on vacation? How about to work? Your answers to those questions show how really important risk factors are to you. And these are real risk factors, with real numbers, not the kind associated with fat and fiber (which I'll explain a little later).

Now, let's say an average two-passenger trip is twenty-five miles; a little more arithmetic predicts that each time you go for a ride you have about one chance in two million of not making it back alive. If you prefer the brighter side, you can claim that 1,999,999 times out of 2 million you will come back alive. How many car trips do you take in a lifetime? Are you starting to count?

If you were paying attention in that last paragraph, maybe you noticed that, on the per-trip basis, planes looked riskier than cars: one chance in a half million versus one in two million. Well, if that car trip were the same as the plane trip, 3,000 miles instead of only twenty-five, the risk would jump to one in 33,000, twenty times higher than the planes, as claimed. The proper comparison is made with passenger miles, not number of trips. You must remember to use the same ruler when you compare two things. The next time you read about the latest study, see if you can tell how many times the writer switches his focus of comparison.

Maybe all this math leaves you cold. In that case, you have more than likely known at least one person who was killed in a car crash. You also have your own concept of all the cars, and all the travel, and all the people, and you read or hear about local fatalities. Believe it or not, you must have some intrinsic feel for how dangerous it is to get in a car. Does that impede you? Do the numbers above change your mind or affect your willingness to jump in? Not likely. So what's the worth of all this risk factor talk?

Risk factors have some real uses. In principal, politicians allocating money for highway safety, hazardous waste cleanup, and other competing life-saving endeavors could use them to direct funds to the more dangerous conditions. But then, politics is probably one of the few areas in which risk factors are consistently ignored.

One would like to believe that federal agencies and similar groups use risk factors in their funding activities, and they no doubt do, in spite of the billions of dollars poured into the Superfund for hazardous waste dump clean-up with often dubious returns. But where does that leave you and me with risk factors? Do you choose olive oil over coconut oil based on a risk factor? Maybe you make the choice thinking that you are minimizing some kind of risk, but that's only because some health guru told you so. It's nothing you really *know* anything about. Chances are that if numerical risk factors were printed on the labels, you would still have to ask "Which is safer?" By the way, do you happen to know how many Fiji Islanders die of heart attacks? I should think they eat a lot of coconut. Then again, what about all those people in the Middle East whose diet started the low-fat kick? Don't they guzzle a lot of palm oil?

I mentioned above that fat and fiber risk factors were not real ones. Let me explain. The probabilities we call risk factors are very small numbers: one chance in two million, one death in a half million. This means that to even make a guess at them, we need a very very large number of examples. An example might be one person getting on a plane; we watch to see if that person gets off alive. You can see that we would have to watch a lot of passengers in order to find that one-in-a-billion who doesn't make it. In medicine and similar fields, we can't wait long enough or don't have enough money to watch enough people to make these guesses, so we cannot determine the risk factors. Not to fret, by a neat little trick it's possible to guess at a "relative risk," instead of a risk factor, by watching only a small number of people. Using this trick for the case of the travelers by plane and car, we could discover that cars were twenty times more deadly than planes, as we did above, *but,* we would never know how deadly either planes or cars are by themselves. So you see, there are two kinds of risk factors: real ones, like the "one chance in two million," and relative ones, like the "twenty times greater." Both these numbers apply to deaths in car crashes. Given a choice, which one would you publish? Have you ever seen a real risk factor published?

Coincidentally, if you were a young American adult in 1992, then according to the *Statistical Abstract of the United States,* your chances of eventually dying from heart disease were about twenty times greater than your chances of dying in a car crash, which, as we saw, was very roughly twenty times more likely than your death in a plane crash.[9] Now, unless you really paid attention, you might run around thinking that car crashes and heart disease are equally dangerous because they are each twenty times worse than something else. This is a trivial example of grounds for some of the raging controversies we follow in the weekly health summaries. Much confusion arises by things being reported out of context even though the headlines are livelier that way.

Here's yet another way to think about risks: If car crashes are twenty times less deadly than heart disease, does that mean that if you drive your car twenty times more than you do now, your chances of dying from either one will be equal? Good question, let's work on it. First, if you're a Manhattan resident, don't own a car, and ride only the subway, then twenty times zero is still zero, so you would have a little difficulty generating a driving record. On the other hand, you, the traveling salesman, spending six hours a day in your car, would have to have six times twenty, or 120 hours in a day to do the experiment! Maybe "twenty times more risky" doesn't apply to you. That's something to think about the next time you're torn in the restaurant between the pasta and beef menus. When it comes to fat, are you like the Manhattan resident or the traveling salesman? To answer the question, even if people drove twenty times more than they do today, assuming the road system could handle it, and if they exercised the same driving skills, that risk factor would not change, even though there would be twenty times more deaths. Yes, that is correct, twenty times more deaths but the same risk factor. Remember, the ruler is passenger miles, which also increased by twenty, leaving the risk, deaths per passenger mile, unchanged.

Just two more points about relative risks that should really help you: You are twenty times more likely to die in a fall than by contracting hepatitis and you are just as likely to commit suicide as you are to die of AIDS (at least as of 1992).[10] Aren't those informative and helpful?

How are risk factors computed? In many cases it's pretty easy. All you need is a count, or guess, of the number of times something happens, say in a week or a year or in twenty years, as for the coast-to-coast flights. This is the number of base events. Notice that neat word, "event"? You will also need a count or estimate of the things in which you are interested that happen in conjunction with the base events. These are often simply called "observations." Formally, the risk factor is the number of observations divided by the number of base events.

As a simple real-life example, the next time you take the kids on a camping trip, keep a count of the number of mosquitoes they successfully swat on their own bodies. Also each day, count the number of new mosquito bites they have. At the end of the trip, the total number of bites is the number of base events, and the count of dead mosquitoes is the number of observations. Let's say there were ten successful swats and a hundred bites. You now have a risk factor: The risk of death for mosquitoes that bite people is (10 over 100) one in ten.

Quite seriously, finding risk factors is basically just that simple. Problems arise when we begin to think too deeply about exactly what we are doing, or when someone starts to examine our report. Nasty details can leave the question open for decades: Were some of the bites actually due to fleas? Did any mosquitoes bite twice? How many? Did the kids truthfully report successful swats? Is that a new bite today or was someone scratching again? Where did that repellent come from?

If this study is important to you, you can see that by the fifth camping trip you will be accompanied by an entomologist, a medical doctor, a recording secretary, one or more videocameras, mosquito proximity alarms, and waterproof skin markers. Your now well-beaten path will be carefully documented as to type of terrain and velocity with which you proceed. It will have designated protected and unprotected rest areas, and be dotted with temperature, humidity, and rainfall recorders. By the end of the tenth trip, when you know that every base has been covered, some joker will argue that the entire environment is now so artificial that you are measuring an irrelevant situation. Now you, and anyone who objects to your research, are in good positions to generate at least twenty or thirty publications in leading journals.

Rule 4: Sample conveniently.

At this point it might be wise to set this book aside until you are in the mood for some serious concentration, say first thing in the morning, or late at night, depending on your biorhythms. Sampling is subtle stuff. Many respectable studies, and their sponsors, have met their doom by not paying attention to sampling details.

What exactly is a sample? One good way of understanding the concept, as well as some of the pitfalls, is by examining the history of a classic failure:

In the midst of the Great Depression, 1936, a Republican named Alfred Landon ran against Franklin D. Roosevelt for the presidency of the United States. Roosevelt was already an American hero for getting the nation started on the road to recovery. Republicans were on nearly everyone's blacklist because, after all, they and their leader, Herbert Hoover, had put us in this mess. No one in his right mind believed for a moment that the Republicans stood a chance in the upcoming election. In spite of the consensus that the Democrats were a shoo-in, someone undertook to make a study. Studies of this type are called *sample surveys*. If they have to do with people, they are commonly referred to as polls. Everyone knows what a poll is. Pollsters contact a number of people across the country and ask their opinion on something. Usually the question is answerable by a yes, no, or I don't know. The votes are tallied and the results broadcast: "Fifty-three percent of gentlemen prefer blondes," or whatever. Sophisticated polls include a footnote admitting that the result is subject to an error, perhaps 4 percent. Of course, no one pays attention to the footnote. It means (almost) in this case, that the true result may be anywhere between 49 and 57 percent. The published number of 53 is just the center of the possible range. If you are male, you may suspect the blonde preference result because they didn't ask you. You *should* be suspect. The reason for the error is the fact that they didn't ask you, or most of the 100-odd million other males in the country. The only way to know the true percentage is in fact to ask every male, and hope that no one changes his mind before the result is published. Obviously, that's too expensive.

Because it's not practical to ask everyone, for your study you must settle for a sample of subjects to ask. In other words, select only a relative few out of the 100-odd million males. If you are clever enough, you can do that in such a manner that their collective opinion will reflect that of all males. How do you do this? That, dear reader, is what makes selecting the sample so difficult. More about that later. Now back to the 1936 election. The pollsters back then faced more problems in selecting the sample than do their counterparts today. There were no massive databases, mailing lists, and so on, that provided names, numbers, and locations of people, broken down by all manner of characteristics. Nor were the pollsters as experienced as today's; they were relative neophytes in sample survey techniques. In any case, faced with the sampling problem and, no doubt, limited funds and time, someone noted that telephone directories were convenient, readily available lists of heads of households in every city and town in the country, and their phone numbers. Voilà! Having made that decision, it's likely that the pollsters applied sophisticated techniques to select the directories used and the numbers to call. (I don't know this, but we can give them the benefit of the doubt, it doesn't matter.) Well, you say, what's wrong with that? What was wrong was the Great Depression. People were so broke that only the very rich still had phones in their homes. Roughly half of all the phones that had been in homes in 1929 had been removed by 1936. And yes, the very rich were mostly Republicans. They wouldn't vote for a Democrat under any condition. The poll made headlines: "Landon Predicted by Wide Margin!" For those who don't recall, Roosevelt's sweep of the nation is still the record. I suspect that the pollsters were on the soup lines by mid-November.

Well, now you know what a sample is. For polls it's specifically who you ask, and how many. For other studies it may be what you measure, e.g., power lines or VCRs, fried or boiled meat, Americans or Japanese. As the 1936 election demonstrated, you had better learn to consider all circumstances concerning your sample, not just the obvious. Would you have connected the content of telephone books to the Depression?

If your answer to the last question was no, don't feel bad. But federal agencies have the resources to hire good statisticians and make extensive surveys. Theirs should be done properly, right? Well, perhaps you remember the study to determine the effectiveness of the war on drugs in the year before the 1992 election. Caution—hot political topic—the result may have been written before the study was made. In any case, workers contacted people by phone, using, you guessed it, phone directories. Everyone has a phone these days so why not? Well, after publication of the (desired?) result—Drug Use Down—case workers, police departments, and treatment centers hastened to point out that inner-city chronic drug abusers frequently have no address, never mind telephones. In this case a retraction was forthcoming, something to the effect that a few key elements may have been overlooked. The lesson here (get out your notebook): Unless you're thoroughly familiar with what you're going to sample, do something that eager researchers forget—make a *pilot study*. A pilot is a small preliminary survey. It provides a lot of practical detail about what to look for and what to measure. Of course, it requires more time and money. However, if you don't do it, chances are excellent that the study you actually do will be worth no more than the pilot you chose not to do. Many reported studies fall in this category. A common problem is that the variability encountered is much greater than was expected and the results are useless.

There is a second aspect of choosing a sample; it was mentioned above but nevertheless is worth expanding: Once you decide on how to select who or what to measure, you need to ask, "How many?" What will be the sample size? To the statistician this is the first question, rather than who or what, for two reasons: The sample size is important in determining how accurate your result will be, and naturally, how much the study will cost.

I note an interesting phenomenon with regard to sample size and published results. Some twenty to thirty years ago, before published studies were fetish, people frequently asked "How many?" The lay reaction to a sample was disbelief. "How can you ask just that many people and claim to know the mind of the nation? Ridiculous." After three decades of intense study-bombing, almost no one wonders anymore. Some sort of mass conditioning (it certainly can't be called mass education) has occurred.

To see how the sample size can get you in the end, suppose you are planning a wedding reception, want to invite many people, and have a grand affair. The rub is, the budget is tight. There is a fixed number of dollars and two contenders for them: number of people and grand affair. Now, you can have all the people and a mediocre house party, or a few people and the grand affair, or some middling compromise. The money for the reception is like the sample size here. Once all the data are taken, there is no way you can have more even though your results demand it. (See, you forgot to do a pilot, underestimated the variability, and so the sample was too small.)

Just like the two contenders for the reception budget, there are two contenders for your sample budget, *accuracy* and *confidence*. I mentioned earlier that the accuracy of the result of a poll depends on the sample size. That's intuitively obvious as well as correct. However, if all goes well, the statistician can make a guess about how accurate the result is, *and* even tell us how confident he or she is with the guess! The accuracy is often in the footnotes, e.g., the survey error was 4 percent. Which means that if you repeated the survey over and over, the results of all of them would agree, within 4 percent in either direction. Almost all of them, that is. Because the statistician cannot be exact in anything, he must allow for exceptions: Perhaps 5 or 10 percent of all the repeat surveys will not be within the 4 percent limit. Some results will differ by more than that. The number of these exceptions he predicts expresses his confidence, or what odds he might give you that the 4 percent margin of error of the next result will include the results of the first study. If the odds are ten to one, he will say he is 90 percent confident. Standard practice is to shoot for odds of twenty to one or a confidence of 95 percent. This means that we should expect (at least) one in twenty studies to be absolutely useless. However, as you might suspect, large samples are usually needed to get good odds. Both the accuracy and the confidence depend on sample size. But relax, you can play games. The error and the odds are like two bungee cords tied together and stretched around a package. If you apply less stress to one, it will shrink while the other stretches. So, for any sample size, after the fact, you can make the error as small as you like at the expense of poorer odds, and vice versa. That neat little footnote is only half the story. Saying that the error is only 4 percent may sound good, but what if, because of the sample size and variability, the odds are 50–50 that it's totally wrong? Just because standard practice is to shoot for a confidence of twenty to one doesn't mean that the authors actually achieved it unless they say so.

Let's look at an important detail of nomenclature: The error, stated as 4 percent, is intended to mean that the margin of error is plus or minus 4 percent around the stated result. The total range of possible results is then ±4, or 8 percent. This range of possible answers is referred to as the *confidence interval.* The 90 or 95 percent confidence, or odds, that we discussed above is called simply the *confidence.* Thus we have, for example, a 95 percent confidence for the 8 percent confidence interval. This is just a taste of the statistician's language. Wait till we get to hypotheses!

If all this sounds complicated, it is. Choosing, and obtaining, a good sample requires that you be something of a seer. Then, absorbing the maze of available statistical methods that might apply (see the following lessons) requires years of study and practice. And, in the end, there can always be the surprise factor you couldn't foresee. So, why bother with all this? Use the convenient 1936 phone book.

Rule 5: Never repeat the study.

It seems that this rule hardly needs stating. Why? Well in general, when you do a study, you're result will be one of two kinds: either something dull, the equivalent of "aspirin relieves pain," or (with luck) a real kicker like my move-north-live-longer nonsense. Now, your first study took a lot of work and money. If the result is ho-hum, you're not about to repeat all that for more uninteresting findings. On the other hand, if you have a fascinating result that's receiving a lot of attention (the media is requesting interviews and professional journals want you to publish more of your findings), there may be an urge to press forward; however, there's that real risk factor that another study might change all of your results. Our first one, or our second one, might be the one in twenty undesirable we just described and the two would seriously clash. Even more realistically, your two studies would clash because both are meaningless; a result of the fact that you don't know what you're doing.

To the conscientious researcher, all results are questionable at best and additional studies may be absolutely demanded by the occurrence of any number of events. Basic to this concern over validity are the confidence level and odds, which we talked about above. After only one study, there is no way to rule out the possibility that the sample you worked with was actually one of the few percent of expected bad ones. This is true no matter how "good" the results appear to be. The choice to go with the result in hand is subjective, there is no rigor whatsoever in such a decision. Stating that the confidence level is 90-x percent is simply a conscience palliative, a measure of your poker game.

Suppose you had determined, with 95 percent confidence, that 53 percent of gentlemen prefer blondes. You might be inclined to accept this. Suppose instead your result were 99 percent prefer redheads. Now what do you do? What would I do? Nothing. I really don't believe anyone seriously gives a hoot about preferred hair color. On the other hand, had the study addressed an issue of health or life, either result should be held suspect until independently verified at least once and preferably more than once. Looked at this way, the decision to publish becomes a moral issue, as well as a subjective gamble. In any case, any result, surprising or otherwise, always demands more work. Besides the chance of a bad sample, there are other reasons a survey can go bad: the variability you found was much greater than you expected, you called after dinner so people weren't interested in food, and so on. Any of these can reduce your results to senseless gibberish, in which case the only way you dare to publish is to totally rely on rule 1, extol or ignore the obvious. Of course, if you didn't realize that it was dinner time on the coast when you called, then perhaps you are not even aware that your result is all wrong. In this case, at least you can publish your nonsense with a clear conscience.

You mean to say that it is useless to perform just one study? This is not far from the truth. Hence the need for rule 5, without which we'll never move forward. To show this, I cite two case histories. In the first, the result was interesting but rather benign. In the second there was ample reason to do more work. Rule 5 was ignored in both cases. See what happened and you judge the consequences.

A few years ago, some researchers interested in the phenomena of left-handedness examined statistics in the *The Baseball Encyclopedia*. As we all know, the population of ballplayers is rich in left-handers. It turned out that, for players listed in the book, the average age at death for lefthanders was nine months less than that for righthanders.[11] This is really a ho-hum kind of result, as well as being nothing more than a peculiarity for the population of players who happen to be mentioned in that book. The observed difference has as much meaning as the number of days since you last trimmed your toenails. Nevertheless, the workers, being true believers in rule 1, published the result. Naturally, the media jumped all over it: "The Good and Lefties Both Die Young." At least two learned articles by other authors quickly followed, attempting to put this "observation" in perspective. But of course, the word was out and no one cared about learned articles; they aren't spicy. But at this point, the researchers had made the spotlight, no real harm had been done, and the whole episode would, in due time, have been forgotten. Did they heed rule 5 and move on? No. They were going to prove their point. They made a second study. This one, fraught with lack of control, demonstrated that the real difference in longevity was not nine months, but nine years![12] That did it. Nobody believed anything they might say. They probably could have gotten out of the mess if they simply pointed out that, according to (both) their studies, lefties will live eight years longer if they just play baseball!

Another good reason for rule 5 is found in the current flap over electric fields. All of this started with an observation by a sharp health scientist in Denver in the early 1970s. It seemed to her that an unusual number of childhood leukemia cases appeared in homes near power lines. Other investigators made a study of this and reported results in 1979.[13] They concluded that "high current configuration" houses had more "case" children. That was their way of saying that there might be something to her observation. For some reason, the media failed to jump on it. However, scientists took note and, with such an alarming result, did the proper thing, they made more studies. At least thirty more have been completed and many are still in progress. In addition, numerous laboratory studies are underway to determine the effects of electric fields on specific glands and tissues under very controlled conditions.[14] This much activity could not be ignored. Results are still incomplete, confusing, and even contradictory. Nevertheless, the media have managed to excite the environmentalists, caused property values to fall, and utilities to be stressed, especially in Canada.[15] In Sweden, utility companies may be forced to reroute distribution facilities. See what can happen when you start making more studies?

In both of these examples doing more studies confused the issue, aroused the media, and excited the public. Obviously, the folks concerned with lefties should have quit after the first study. But what about the electric fields? The issue seems important, but look at the penalties already incurred. Where is the fault or blame there? Should rule 5 have been followed? As I said, you be the judge.

Notes

1. Tavia Gordon, "Further Mortality Experience among Japanese Americans," *Public Health Reports* 82, no. 11 (November 1967): 973–84.

2. Ancel Keys, "Epidemiologic Aspects of Coronary Artery Disease," *Journal of Chronic Diseases* 6, no. 5 (November 1957): 552–59.

3. Stephen B. Hulley et al., "Health Policy on Blood Cholesterol, Time to Change Directions," *Circulation* 86, no. 3 (September 1992): 1026–29.

4. "Castelli Speaks from the Heart," *AARP Bulletin* 33, no. 5 (May 1992): 16.

5. Robert V. Levine, "The Pace of Life," *American Scientist* 78, no. 5 (September–October 1990): 450–59.

6. U.S. Department of Commerce, Bureau of the Census, "Death Rates by Selected Causes—Selected Countries," Table 1366, *Statistical Abstract of the United States 1992* (Washington, D.C.: Government Printing Office, 1992).

7. U.S. Department of Commerce, Bureau of the Census, "Worldwide Airline Fatalities, 1970 to 1990," Table 1042, *Statistical Abstract of the United States 1992* (Washington, D.C.: Government Printing Office, 1992).

8. U.S. Department of Commerce, Bureau of the Census, "Motor Vehicle Registrations, 1960 to 1990, Vehicle Miles of Travel, 1990, and Drivers Licenses, 1990, by State," Table 1001; and "Motor Vehicle Accidents—Number and Deaths: 1970 to 1990," Table 1008, *Statistical Abstract of the United States 1992* (Washington, D.C.: Government Printing Office, 1992).

9. U.S. Department of Commerce, Bureau of the Census, "Deaths and Death Rates by Selected Causes: 1970 to 1993," Table 125; "Acquired Immunodeficiency Syndrome (AIDS) Deaths, by Selected Characteristics: 1982 to 1994," Table 130; and "Deaths and Death Rates from Accidents, by Type: 1970 to 1992," Table 134, *Statistical Abstract of the United States 1995* (Washington, D.C.: Government Printing Office, 1995).

10. Ibid.

11. Diane F. Halpern and Stanley Coren, "Do Right-Handers Live Longer?" *Nature* 333 (1988): 213.

12. Diane F. Halpern and Stanley Coren, "Left-Handedness: A Marker for Decreased Survival Fitness," *Psychological Bulletin* 109, no. 1 (1991): 90–106.

13. Karen Fitzgerald et al., "60 Hertz and the Human Body, Parts 1, 2, 3," *IEEE Spectrum* 27, no. 8 (August 1990): 23–35.

14. Edward Leeper and Nancy Wertheimer, "Electrical Wiring Configurations and Childhood Cancer," *American Journal of Epidemiology* 109 (1979): 273–84; and Tekla S. Perry, "Today's View of Magnetic Fields," *IEEE Spectrum* 31, no. 12 (December 1994): 14–23.

15. Leeper and Wertheimer, "Electrical Wiring."

Lesson 2

The Nitty-Gritty

Now that you know the rules of the game, you must learn the methodology. Brace yourself, here comes the technical stuff.

All technical subjects have one thing in common, a coveted collection of nouns and verbs that you seldom find in *Webster's Dictionary*. I suspect that large sums of money are devoted to keeping them out. Why? Simply because, when you have the definitions, you, too, know what's going on. The premise is even true for the so-called hard sciences, physics, chemistry, and the like: Once you know the jargon, you can comprehend and even converse. This has been amply demonstrated over the years with serious TV documentaries exploring hitherto esoteric subjects. They are often successful in bringing science to the homestead, albeit just for a day. A key to that success is a liberal use of everyday English and avoidance of the technical tongue-twisters. Even the "queen of sciences"—mathematics—is reducible to amusement and low-level participation when translated into basic English. There are a number of popular books on math-oriented games and puzzles, presented for the most part without the mathematical lingo. Your first task in mastering the nitty-gritty is, therefore, to build your vocabulary. The methodology will fold in as we go along.

From lesson 1, you already know about *correlation, sampling,* and *confidence level,* and that knowledge alone raises you above the crowd. To become even better, there are a few more nouns and verbs you should have at your fingertips.

We begin with the simplest and most pervasive of all statistics, the average. Everyone knows what an average is. A friend of mine likes to put it this way: I'm just lying here with my feet in the freezer and my head in the oven. On average, I'm quite comfortable.

Golfers know their average distance with a two-wood, and you know what your average take-home pay is (in dollars, not how you feel about it). We encounter and talk about averages every day. As a matter of fact, statisticians seldom do anything except calculate averages. Their basic texts are little more than dictionary-like cookbooks, wherein each recipe for a new average is assigned a new name. You will see some of this as we move along in this lesson. Another word of caution, never, never refer to any of these things as averages: They are means, or, for experts only, *expected values* or *expectations.*

On this subject of averages, there is a popular trap you can fall into, especially if you're a baseball fan. The batting average in that game is not a simple average at all. Batting 300 means that, at a thousand times at bat, the player is expected to get a hit 300 times. The 300 is the ratio of hits to the magic 1,000 ups, written somewhat like a percentage. It's just that the fraction is multiplied by 1,000 instead of 100 and, in this case, the ratio 300/1,000 becomes simply 300. So don't give yourself away by discussing Dimaggio's batting average outside of your circle of friends. Talk about his "expected proportion of successes." You might even add, "in a thousand trials." That should clinch your credentials.

Now, you golfers, when discussing the length of your drive with a two-wood, you may say "200 yards, give or take five." It is then unlikely that your partner would comment if one fell anywhere from, say, 190 to 210 yards away. If it dropped at 185 or less, or 215 or more, you might hasten to explain the problem or say "lucky hit," as appropriate. Statisticians and researchers can't afford that sort of sloppy communication. The five yards of your "give or take five," is the *standard deviation* of the *distribution* of your drives! As promised, this new statistic, standard deviation, is just another average. In this case it's the average variation of your drives around the average distance of 200 yards. Standard deviation is so important that, in a pure form, it has a special name and symbol, the Greek letter σ (*sigma*).

Let's take a breath and pause to digest this (I told you to brace yourself.)

Three new names have been introduced in the golfing paragraph: standard deviation, distribution, and sigma. I prefer the simple sigma rather than standard deviation, so I will ignore some impractical differences and not use the latter any more. You ought to remember it though. Now, distributions and sigmas go hand in hand. Let me explain:

You and your golf partner know and understand that all your two-wood drives will not be exactly 200 yards. The only reason you mentioned the give or take five was to give the guy an idea of how consistent you are. As you will see, that conveyed a lot of information that the statistician formalizes. Your, and your partner's, comments on any particular drive are based on all the data you passed on, almost unconsciously, with "give or take five."

The lack of comments for drives between 190 and 210 yards means simply that those were not unexpected. They were reasonably close to your 200 yard "average." The statistician quantifies that word "reasonably" by (essentially) observing that in practically all real cases, most of the measurements in a whole group fall within plus or minus two sigma of the mean. These measurements are therefore not uncommon and so are declared to be reasonable. That is, most of your two-wood drives should fall between 190 and 210 yards, or within ten (two times five) of your average. The number, five, is the sigma you passed on with your "give or take five" statement.

Begin to get the picture? The average value of something may be of interest, e.g., "What was the average Dow Jones value last year?" However, if you play the stock market you are painfully aware that there is a lot more to stock prices than the average. The sigma of last year's Dow Jones would give you an idea of how much it fluctuated. Most of the daily Dow Jones numbers would have been within two sigma of that average.

In practically all things, usual events are within two sigma of the mean. On the other hand, any event that is plus or minus three sigma or more of the mean is, at least, worthy of note. That lucky two-wood hit of over 215 yards is an example; the student statistician might exclaim, "Wow, a three sigma clout!"

All of these things are formalized not merely by putting names on them but numbers as well. Strictly speaking, with a sigma of five, your chances of hitting the ball more than 215 yards (the mean plus three sigma) are about one in a thousand. Real world record setters, the one in a million type, do things at the four and three-quarters sigma point. Isn't it neat to have all that meaty terminology at your finger tips? Forget that mundane "200 give or take five." Unless you do, you'll never make it in the studies circle.

Before leaving this sigma business, we must mention another

closely related average known as the *variance*. Variance is really redundant because it is nothing more than sigma multiplied by itself, or squared. (Because of the form of the calculation, variance is the first result, and then sigma is the square root of variance. Never let it be said that we are not precise.) In practice, sigma is more useful than variance because it directly measures the chances of very common or very rare events, the common ones are within two sigma, the rare ones greater than three sigma. However, use variance when you want to emphasize what it sounds like, variability. Why? Well, big variability makes for big problems. Remember, the prime function of the statistician is to make estimates, and large variances make that harder. For example, if you shoot at a target with a very good rifle and you're a good shot, the target will have a bunch of holes clustered near the bullseye. That tight cluster has good *precision* or small variance. (Mathematically, precision is just variance upside-down, the reciprocal.) If you're a lousy shot or the weapon is bad, the holes will be all over the target; there is little precision and there is large variance. In the case of the bad gun, making guesses or estimates about where the next shot will fall is obviously difficult and pretty risky. So, brag about big variance when you want to impress your colleagues with how tough your study was. With a small variance you can emphasize precision and consistency, e.g., "everyone (without exception or variability) who ever ate bacon and eggs during any part of the eighteenth century is now dead."

As you might have guessed, the confidence intervals we talked about earlier are strongly affected by variance or its cousin, sigma. Large variance makes for poor confidence and requires larger samples for usable results. You can get a feel for why larger samples are needed by considering the targets used with the good gun and the lousy gun. It would be easy to estimate the center of the cluster of holes on the target with the small variance, but doing that for the other target would be tricky unless there were a lot of shots fired, creating lots of holes or a large sample. Finally, suppose you let a friend use that lousy gun and he claims he put nine out of ten shots right in the bullseye. Would you believe him? The lesson here is to look for large variance in the data, then you can judge the worth of the report. Caution, large variance isn't always as obvious as the scattering of bullet holes all over the target.

For example, suppose you read about a study in which 10 percent of people with a certain disease who took preparation Y died. A typical news report might stop at that. Now, it may be that another

10 percent became more ill, 30 percent showed no effects, and 50 percent recovered from the disease. This shows a lot of variance in the results and you should be suspicious. How do you know that those 10 percent died from preparation Y? How do you know that the 50 percent recovered because of it? And, why was there so much difference in the reactions? Until those questions are answered, the study has no business being reported except in the medical journals, and there as an informational item only.

The remaining new term, *distribution,* is fairly self-explanatory. A distribution gives a picture (you can draw it) of all the different distances for your two-wood drives, or how the bullets splattered around the target for either the good gun or the bad one. There are all kinds of distributions, and the nomenclature here is totally out of control. You may run across names like normal, chi-square, student's t, random, uniform, and on and on. Don't panic when you see yet another new name. There is no need to study dozens of them. As you will see below, there are mathematical facts and researcher fantasies that simplify the whole thing. All one really needs to know about is the good old normal one, which by the way, may assume other names at the discretion of the user: standard, bell-shaped, and for the purist, Gaussian. (Mr. K. F. Gauss, a German mathematician, had a lot to do with it.)

The arithmetic dealing with most distributions gets to be downright voluminous, and is strictly relegated to computers these days. However, as I said, the only distribution one needs to learn something about is the normal one, for three reasons: First, it seems to be a quirk of nature that just about anything of interest is found to be "normally distributed." To know what this means, think about the height of all the people you know. Better yet, use your imagination and think of the heights of all adults in the nation (don't forget dwarfs and basketball players). Now, if you can, imagine them all lined up, side by side, with the tallest in the middle, then sloping down to one each of the two shortest placed at the ends. Step way back so you can see them all at once. Is the outline sort of bell shaped? Maybe a little squashed, but generally like a bell? There, now you know why the name "bell-shaped" is used. This is the general idea, but the actual distribution of heights of people is a little more subtle. Let's play a game. It may seem a little rough, but this is only in the imagination.

Tell all the people who have the same heights to form separate groups and lie down. Now, this is where it gets rough. Start with the group of shortest people and stack all those bodies up, one on top of the other, maintaining the prone positions. This is easy, there are, at most, only two of the very shortest, one from either end of the long line. Take the group of next shortest and do the same thing, right next to the first group. There still may be only two or three. Continue with the next shortest, and so on. By the time you get to the group of average height you will probably need to use a long ladder to help with the stacking, the pile will be rather high. However, at the end, there will only be one or two of the very tallest folk to stack up last. So, now you have short stacks of the few short people at one end, short stacks of the tallest people at the other, and very high stacks of all middling height folk in the center. If you followed all that, imagine stepping way back to see the outline of the stacks of feet facing you. This will really be a bell-shaped curve and is, in fact, a normal distribution (of heights of people, not of feet). The difference between this outline and the first one when all the people were lined up is that now the average height people are all exactly in the center. All those of less than average height lie to one side and all those taller than average lie on the other. At one extreme is the shortest person in the country, at the other, the tallest. I don't know the exact average height of Americans today, but let's say it is six feet, just for convenience. Similarly, say the sigma is about a half a foot, then the mean plus or minus two sigma is six feet, plus or minus one, for a range of five- to seven-feet tall. If you had to wash the feet of all those (within) two sigma folks lying in the stacks of people five- to seven-feet tall, you would wash 95 percent of all the feet in the country. (If no one had suffered an amputation, that is; remember, we have to assume some things.) It would be much easier to scrub only those that lay in the extremes, beyond the three sigma points; that is, people less than four-and-one-half- and more than seven-and-one-half-feet tall. The total beyond those points is only a quarter of 1 percent of all the (adult) feet in the country. That last number, one-quarter of 1 percent, is so small that it is common practice to ignore those folks, known as the extremes, completely, unless you're strongly pro- or anti- something or other. In that case you must make a big deal out of them. The extremes can always be relied on to complicate things because they include all the exceptions, the odd cases, the situations that belie what the averages seem to show. If you ever make studies of your own, take heed: When you write your report, you had better explain those extremes satisfactorily, or else be prepared to defend yourself against the extremists.

As I suggested, if you play games like this with almost anything natural—peoples' heights, the times you arrive at work, the dollars in your wallet from day to day, the speeds of cars on a highway, whatever—you will almost always get a normal distribution. That is the first reason for its importance. The other two reasons are related to each other and are just slightly technical.

Recall that the arithmetic for most distributions, e.g., calculating actual percentages for a given number of sigmas, is tedious at best. The normal one is no exception. Showing that 95 percent of all those feet are within two sigma of the average is a nontrivial task. However, because the normal distribution appears all over the place, statisticians bit the bullet a long time ago and laboriously cranked out all the numbers anyone would care to use concerning it and placed them in cherished volumes called Probability Tables. Your local library should have several of them.

Probability tables contain pairs of columns of numbers. One of the columns may be abbreviated in some way to save space, but you can always match up a number in one column with its corresponding number in the other. The numbers in one column will correspond to the thing in which you are interested, e.g., heights of people in the United States. The other column will give you the expected fraction of people who are that tall or shorter. If, for instance, you look up six feet (and that is truly the average in America) then you will see the corresponding number 0.50. This tells you that 50 percent of all the people are six feet or less in height. Similarly, if you looked up seven feet, the corresponding number should be 0.98, meaning that 98 percent of all of us are that tall or shorter. Note that those percentages also correspond to the *probability* of finding a person in that height category.

There is a minor hitch. Because the tables are meant to be universal, that is, apply to all things that are normally distributed, the first column, in which you located the height of interest, is actually given in terms of sigmas. Thus, the average value, instead of listed at six feet, is listed at zero sigma and seven feet would correspond to two sigma. The tables are used to look up the probability (of being at 1.3 sigma, for example) and you must, somehow, determine what your own mean and sigma values are. This is the only way one table could work for cases with means of either 100 or 1,000, for example. There is no way one could write down tables for every possible combination of means and sigma values.

Having done that, and not wishing to repeat that monumental

task, the mathematicians and statisticians spent and still spend great effort to find tricks to convert other interesting (or not so interesting) distributions in such a way that they look like the normal one. These tricks are called *transformations*. Even though they are actually called tricks, they are in fact legitimate mathematical operations. The beauty here lies in the fact that, once a transformation is found and then used to convert some funny-looking distribution into a normal one, it can be used in reverse to translate normal probabilities, read from the tables, into meaningful values. For a silly example, imagine that people were not normally distributed in height. Suppose that tall people were very much taller than they are, say, 100 times taller. In addition, people just slightly above average were stretched so they are ten times taller. Let average people remain average but, shrink the short people and, the shorter they are, the more you shrink them so that the shortest people are made to be 100 times smaller. Now think about that distribution of heights, with the people laid out somehow on a very large ruler. There will be a big pile of prone bodies at the very small values of height and those tall folk will be spread out very thinly over the range of tens to hundreds of feet. We call that a *skew* distribution and, in this case, it is the logarithm of height that will be normal, not the actual height. To use the tables we would have to convert some height, say six feet two or 74 inches into the logarithm of 74, 1.87, and look up that number. The probability found for 1.87 would be correct for the (reverse transform) height of 74 inches. So, we don't have to do all the arithmetic to generate a table for numbers that are normal on a logarithm scale, what is known as a log-normal distribution. Note that some distributions are so strange that it is actually impossible to truly calculate the probability tables for them.

The widespread practice of using transformations to make things look normal is the second reason that the normal distribution is the only important one for our purpose.

The third reason for the importance of the normal distribution is even more pragmatic. When a trick cannot be found to make some distribution look normal, or if it defies description, it is common to say, "Well, it is nearly normal," or more boldly, "Assuming normality . . ." and just plow ahead. If in fact the distribution is not normal, the odds are good that much of what follows will be (even statistically) invalid. Technical literature of all kinds is loaded with those phrases simply because the real distribution is very difficult to determine or even unknown. Watch out for such phrases if you ever have to read a real report. The author may have been able to justify them or they may be garbage. Of course, in your own report you would use such words with discretion and only if you could justify them.

By the way, if you do write a report, you will certainly talk about what you were measuring, be it the height of people or holes in targets, whatever it was you proposed to study. Vocabulary is important in order to sound more technical and confusing. Except in the title of the report, specific nouns are taboo. To appear to be an expert you must converse in generalities so as to be all-inclusive. The things you measure are not feet or holes but *random variables,* which is somewhat self-explanatory at a gut level. After all, if they weren't variable, you would only have to measure one. For simplicity, they may also be loosely referred to as *variates* or *deviates* (the latter becoming archaic but still acceptable even in polite society). Thus, if you see a table of numbers called "Values of Standard Normal Deviates," there really isn't a conflict in terms. It's perfectly okay to talk about deviates that appear to be normal or even uniform. However, until shown to be otherwise, deviates must be considered as random. Later, we may talk more about what "random" really means. For now, your intuitive "sort of all mixed up" definition will suffice.

Since beginning this lesson, you have learned more than twenty terms from the dark pages of the statistician's glossary. Maybe you don't realize it, but you have already become something of a practitioner. No longer will you dial 911 when you see a non-Gaussian deviate being transformed! You might even inquire after the variate's expectations.

Since you're really moving along on the road to expertise, it's appropriate to take some time here to emphasize a few important points. Some of these may seem obvious, but, in your future readings, note how often they appear to be overlooked (or perhaps wantonly ignored).

I stated earlier that many studies begin with correlations. Now human beings, or most of us, naturally believe in cause and effect (i.e., there must be a reason for everything that happens), and expect the world to behave that way. We tend to get upset when it doesn't, and may even entertain irrational thoughts. Recall the enlightened Neanderthal measuring darkness and temperature. If there were never any warm winter days or cold summer days, his data would have indicated true cause and effect, or at least a common cause. However, those anomalous warm winter and cold summer days complicate things. The witch doctor could handle them by simply telling his people to dance harder around the stone idol, but they force us to use fuzzy terms like correlation that subtly hide what we don't wish to face. We'll see a lot of that sort of thing in lesson 6.

For now, I just want you to see how this might work. Suppose archaeologists dug up some of the Neanderthal's data. In the absence of clocks and thermometers, his entries were simple pairs of symbols describing lengths of nights and next-day temperatures, e.g., long–hot, short–cold, and so on. In the diggings it is found that 100 entries survived the eons of burial. Their summary: 45 short–hot (summer?), 40 long–cold (winter?), 8 short–cold, and 7 long–hot (what?). Eighty-five percent of his data shows a simple relation between darkness and temperature but, 15 percent of it says *no way!* There is no clear-cut answer. The data are contradictory and appear to be problematic. However, the statistician or researcher can sweep these 15 percent under the rug. Because those 15 percent present the kind of situation that bothers people, we would like to ignore them, or at least minimize or even eliminate their obvious negative nature. There is a strong temptation to say something like "Usually that's the case," or "Most of the time," or something of the kind. The sophisticated researcher doesn't need to do this. Without so much as a second thought, he turns to the trusty statistician and asks for the correlation coefficient, which must be a number between –1 and +1. Don't worry about how to compute it, that involves a lot of arithmetic. If there were almost no contradictory data, say only one or two days of short–cold recordings, the number would be close to +1, perhaps larger than 0.8, and both readings, or values, would go together in the same "direction," e.g., longer nights and warmer days or shorter nights and colder days, if we think of long and warm as the same direction. Conversely, if there were few short–cold recordings and we keep our notion of direction but short nights appear to predict warm days, then the coefficient will be close to –1, perhaps smaller than –0.8. Finally, if there is a large amount of conflicting data, so much that it overwhelms the "good" recordings, the coefficient will be near zero. In statistics one should never get coefficients of exactly plus or minus one. That would "prove" that the thing you studied was not statistical. (A trivial exception to "never" occurs if you correlate a variable with itself. We will show some examples in later lessons.) So, correlations near plus or minus one indicate little contradictory data and, the closer they get to zero, the more confusing the pictures become.

In this case the coefficient turns out to be +0.7, a number big enough to make any study-monger jump for joy. (They get interested if it is only 0.1 and swoon at 0.5.) The researcher may be in ecstasy for days over this one and never give a thought to the contradictory data points. The message is this: Whenever you see a published correlation coefficient, don't be lulled into thinking only of the data that support the thesis (85 percent here), but ask how much of the data refute it! For our witch doctor friend, 15 percent of his data (100–85) contradicted his summary statement. As a matter of fact, if you can get a copy of the data, you can scream and holler that x percent contradict whatever it is the report is claiming. The next time you read about a study reporting that p percent did whatever, remind yourself that 100 minus p percent did the opposite. That simple trick often reveals a much more interesting result. One good example is a 1993 report published in the *New England Journal of Medicine* that claimed that 4 percent of all heart attacks were triggered by overexertion. Naturally, it went on to rant about how important it is to avoid that sort of thing. It never mentioned that, therefore, 96 percent of heart attacks had nothing to do with overexertion. Nor did it consider the very likely possibility that, in all things that men do, they may be caught overexerting at least 4 percent of the time and so some 4 percent happen to be doing that when their attack occurs.

I said that the correlation of 0.7 was something a researcher could get excited over. An example may give a little more feel for just what that means. If the Neanderthal began to predict next day temperature based on the length of darkness, and if the pattern above continued, he would be right 85 percent of the time. (Note, however, that a correlation of 0.7 doesn't always mean 85 percent.) This, by the way, is about as good as modern weather forecasters do, and at least four times better than say, a prediction of death by cancer based on smoking. Why is it then that we love to attack the weatherman but find the purveyors of risk so fascinating?

To continue with our new vocabulary, more unique terms arise in studies involving testing. In the more scholarly reports, phrases such as *control group, double-blind,* and *expected proportion* are common. These become self-explanatory when some of the problems intrinsic to such studies are recognized.

One of the major problems in testing is a phenomenon discovered in the early part of this century by managers trying to increase productivity. At a Western Electric plant known as the Hawthorne

Works, someone wondered if better lighting would enhance production. It was decided to test the idea using a group of workers in one location in the factory. New lights were installed, increasing the illumination. Yes! Productivity jumped. If a little light was good, maybe more light would be even better, so more lighting was installed. Yes! Productivity jumped. This process continued until the lighting level was in danger of embarrassing the sun, at which point someone became suspect. The lighting level was reduced significantly and, yes! productivity jumped. Finally, the lighting was put back to the initial conditions and, you guessed it, productivity jumped. The lesson was simple, just paying attention to the workers apparently made them feel good and they worked better. The phenomenon is generalized to mean that whenever people know they are being treated in a special manner, their behavior will not be normal. Naturally, it is called the *Hawthorne Effect.*

Now, suppose you wanted to find out if cough syrup would help mend a broken leg. You understand that giving the syrup to just one or two people with broken limbs would not be very scientific. So, among the first things you need is a large supply of broken legs and a reliable source of cough syrup. A moment's reflection will tell you that if you treat all of the people with the syrup, and find the time it takes to mend the average leg, you won't know much because you have nothing with which to compare your result. You won't know if the cough syrup helped. This is where the control group comes in. For half of the cases of broken legs, you do nothing special, that is, don't use the syrup. The idea is, if the syrup helps, this control group of broken limbs should take longer to heal than your test group, the other half of the cases to whom you administer syrup. That sounds pretty neat but there can be real problems controlling the control group. First of all, there is a danger that the Hawthorne Effect will mask normal bone healing. Well, you say, we won't tell those folks that we are watching how fast they mend. That's a good strategy if you can truly achieve it, people get suspicious very easily. A second strategy is to use the effect in a positive way. Give some syrup to both the test and control groups, but make the control group syrup just flavored water, a *placebo.* The idea here of course, is that one expects both groups to heal at a faster-than-normal rate because of the Hawthorne Effect or similar phenomena, but any difference between the two groups, if significant, must be due to the good syrup.

The control group and placebo schemes help but are still not foolproof. You, yourself, and the people administering the study and the syrups, cannot be trusted. It's an accepted fact that you (collectively) can easily, even subconsciously, bias the data. If you do everything yourself, you will tend to make the desired result come true. For example, in doling out the medicines, you might pour a little extra of the "good" stuff. Or, in deciding when a leg is healed, you might be a little optimistic for the study cases and pessimistic for the placebos. Such actions are usually unintentional and unknown, even to yourself. Don't tell me that you're above that, I won't believe it, and besides it's something nobody could prove, either way. In order to make your results more believable you must resort to the *double-blind* technique. That's a catchy name for a system wherein the subjects with broken legs don't know which kind of syrup they are getting and, in addition, the people doling out the syrup don't know which is which. To do this, someone must make up a master list recording things like "subject 73 gets syrup X, 74 gets syrup Y," and so forth. Whether X or Y is the placebo must remain a secret until the bitter end.

Now the whole thing is foolproof, right? Wrong! Did you look closely at which people were in each of the two groups? It is possible that the control group is heavily weighted with people who have brown eyes and can you prove beyond the shadow of a doubt that brown-eyed people don't heal faster or slower than others? Suppose one group contained a majority of Germans or Chinese or anything else. Can you always be sure Mother Nature is not going to shade your test? You must have identical groups.

Before you object that identical groupings of people are impossible, consider two kinds of identity: *homogeneous* and *heterogeneous*. The first is the obvious one which could be achieved only by using dozens or hundreds of identical twins and placing one of each pair in each group. That would probably be okay, but trying to get them all to break their legs to accommodate your study might be tricky. The alternative is the identically heterogeneous groupings. The concept is to recognize that all folks are different and then to make both groups very large, in the sampling sense, and from the same general population. In this way you hope to get the same degree of variability (of all kinds) within both groups. The plan is then, that if there are any characteristic biases, such as brown eyes, they will occur equally in both groups and wash out. Yes, that's the idea and the hope. Lots of luck in achieving it. Opposing research groups may argue for years over this point. Proving that you did not achieve similar heterogeneity is about as difficult for your adversary as it is for you to prove that you did. Any discussions of this question after the fact must almost always result in a draw. A real-life example of this, in a study of considerable importance regarding the effectiveness of heart transplants, appears in lesson 7. The major point of contention, which was never resolved, centered around suspected biases in the control group. As you will see, the discourse continued for years. Occasionally, years later, some other study or finding may uncover a previously unknown factor that proves your groups were in fact biased. However, happening years later, chances are no one will even care.

Now, having run out of cough syrup, your study is complete and it's time to examine results. There are at least two ways of summarizing the test. The two methods may give consistent or conflicting results. However, you can do what it appears that some researchers do, use both and, in case of conflict, report the one you like. If they agree, use both and you have a double-barreled argument supporting your study. In the first case, simply find the average time, in each group, it took for the bones to heal. In the second case, find the time it takes for the median number (one-half or 50 percent) of the broken bones to heal; again, a separate number for each group. Both of those methods are perfectly valid and, on the surface, might seem to be the same. But, consider the following.

There are many factors that tend to make these two answers different. First, because people are all different, they take different amounts of time to heal from similar injuries. One person's broken leg may be in great shape after four weeks, but another's may take seven or eight weeks to reach the same state of repair. As a result, the time required in each group for the median number of bones to heal will probably not be the same, perhaps five weeks for one and eight for the other. As so often happens in actual studies, the real reasons will likely have absolutely nothing to do with your cough syrup, but of course you won't concern yourself with them because your test was never designed to look for them. Besides, the likelihood is vanishingly small that you could ever find them.

To further complicate matters, in the group that had mostly fast healers, there could be a few very slow healers—say of the six-month variety. Those few don't affect the one-half who healed in five weeks, but they raise the devil with the average. They could very easily make their group's average healing time much longer than that for the group that healed slowly according to the median measure. This sort of apparent conflict is not at all uncommon. Nor is it uncommon, intentionally or otherwise, to report only the particular result that one deems favorable.

Whether the average healing time decreases or 50 percent of subjects heal faster is not very significant. If the two results differ, it must be due to the behavior of those very slow healers and one should take a close look at them. In either case (of improvement), it seems that the cough syrup helped some of the subjects but there is certainly something of interest in the slow healer category and those cases should be examined.

One way of looking at the impact of slow healers on the value of your cough syrup is to separate them from the rest so they can be examined as a group. An approach to this is to look at fractions other than the 50 percent as above. Suppose that 10 percent of the group comprised slow healers but we don't know that at the start. The test results can be looked at on a daily basis and we might see that 10 percent healed in three weeks, 30 percent in four weeks, 50 percent in five weeks, etc. If this were done for both groups we would have a much clearer picture of just how much improvement was achieved, for all types of healers, slow and fast. Note that this would also put those very slow healers in proper perspective, allowing them to be evaluated separately and not confuse the issue.

The name for this kind of analysis is *expected proportions.* That is, we determined the proportion of subjects expected to heal within so many weeks. For instance, we may have found that 30 percent of the test subjects, those taking the syrup, healed in four weeks but only 20 percent of the control group healed in that time, a difference of 10 percent. We might then claim that 10 percent of the test subjects healed noticeably faster than the control group.

A slight variation in this approach compares the times to heal a fixed proportion instead of proportions that healed in a fixed time as we just did. Rather than looking at the four-week interval and the corresponding 30 and 20 percent numbers, we wait until 30 percent of the control group healed, say five weeks. We might then claim that the syrup cut healing time by 20 percent (five weeks to four weeks) for 10 percent of the subjects. As you see, this is just a little different way of saying almost the same thing. Both statements, from the same study, are correct, but the second one probably carries more impact. The media, or perhaps your grant committee, would be more impressed. Anyone truly interested would have to look at your data to really understand either statement. If you do a report, and you want to ensure next year's grant, play with both methods of presentation. Depending upon context and particular results, one may serve your purpose better than the other. You may even choose to emphasize the minority situation. For example, for the hair color survey (remember the blondes and redheads?), would you have advertised that 53 percent of gentlemen prefer blondes, or that 47 percent don't care for them? Putting the emphasis on another aspect may switch the entire impact. This is often done in reporting pre-election public opinion polls. Each political party puts the headline emphasis on the aspect deemed most favorable. Try thinking this way when you read tomorrow's health column.

Now, let's get back to vocabulary. Specifically, we need to learn about *hypotheses*. When discussing hypotheses, we will often talk about events having *probabilities* of occurring. It's an awkward word that sometimes smacks of scientific snobbishness. It's hard to spell, so I'm going to use the word "chance" instead. I did that earlier but now I want to make it clear that from here on out, chance means probability (almost always*). Further, chance may be expressed as a decimal number (0.1), a fraction ($\frac{1}{10}$), or a percentage (10 percent), all the same thing.

Now, about hypotheses: Statisticians, and researchers who use hypotheses (or other statisticians' tools), seldom resort to plain English to describe what they're doing. In our study of bone-mending cough syrup for example, we would never say simply that we were measuring the difference in healing times for those drinking syrup and those who did not. Instead, we would discuss our hypothesis that "the difference in mean times to mend is significant." Our data analysis will test the hypothesis and cause us to accept or reject it, at a significance level of (say) 1 percent. Or, if it is preferred, we can get even more elegant and precise, and claim that at the 1 percent level our data rejected the *null hypothesis,* that there is no difference. Which means, approximately, that cough syrup seems to help in 99 (100 minus 1) out of 100 tries.

Hypotheses are fundamental to the "verification" of risk factors, in particular, those unreal relative risk factors we mentioned earlier.

---

*In the realm of probability, practitioners never speak of absolutes. Even a probability of 1.0 (or zero) does not guarantee that the event will (or will not) occur. Expressions such as "almost always" or, "with probability one," appear throughout the literature. Thus, unlike physics, an unexpected event does not mean that something is wrong.

Recall that they are the ones that can really scare you with claims like cars are twenty times more deadly than planes. Since these relative risks are the ratios of two real risk factors, the usual null hypothesis is "There is no difference," or, "The risks are the same," or, "The relative risk is one." Remember that the null hypothesis is usually a proposition that we hope is not true. Let's suppose that we calculated a relative risk of 3.0 for eggs dropped on a hardwood floor compared to eggs dropped on a carpeted floor. This means that in our test, three times as many eggs broke on the hardwood than on the carpet. In order to be objective we might then ask: Is three a reasonable number for this relative danger? One way of testing for that is to rephrase the question to "What are the chances that, in my experiment of egg dropping, the result of three as the relative risk is coincidence (occurred by chance)? If that were the case, then my experiment would be useless and I could not make any statement about risks. The answer has to do with something called *significance level.* Now, wouldn't you think that a high significance level would be desirable? Aren't significant things usually outstanding and large? If you think so, then you haven't yet developed the statistically shaped head. The significance level is the chance that the number three came up by accident and that the unwanted null hypothesis is true after all. Therefore, you hope that the significance level is insignificant! You will find proudly published significance levels ranging from 0.05 (5 percent) to really great ones like 0.000001. In lesson 6 I'll show one even smaller than that having to do with alcohol and heart disease.

Now each time we test an hypothesis, we come up with some number for the significance level. That number is the chance that there is actually no difference between the two things tested even though we *happened* to get different results. In our case of egg dropping, it is the chance that there is no difference between the hardwood and carpeted floors. So, what is a "good" significance level? It is whatever you think is good, or one that you can talk your colleagues into believing to be good. Some researchers will settle for 0.10, some do not become interested unless it is smaller than 0.01. If such numbers really came from our test then, in the first case, one would expect that in one trial out of ten, the same number of eggs would break on either floor (or more would break on the carpet) and, in the second case, that would only happen in one trial in 100. Whatever significance level you settle for is your choice, it is subjective. Now, when you read in the columns that A is thirty-three times more risky than B, how often do you see the significance level? Without that, you really have no information.

Not surprisingly, there is another popular name for the significance level which seems to derive from the fact that it stands for the probability that you are wrong in whatever it is you conclude. This other name is simply the *P-value.* Almost any study report (not the newspaper columns) will include tables of risk factors, each with its corresponding P-value or, sometimes just *P.* I, however, prefer to call it what it really is, a *confusion index.* The closer the confusion index gets to zero the more believable your results become. This is much more sensible because, as a researcher, you are trying to reduce confusion, aren't you? I would not abbreviate the confusion index as C.I. of course, as that would confuse it with confidence interval, which you will remember was discussed in rule 4 of lesson 1.

If this hypothesis stuff seems a little heavy, don't worry. At least you have been introduced to the terms, which is more than most people can say. As a rule, you needn't get deeply involved. Occasionally however, you may run into the type of person who has a real need to drop terms or ask questions, just to impress. In the hypothesis formulation and testing business, this character is apt to be wondering about type 1 and type 2 errors, also known as errors of the first kind and errors of the second kind. (No, there is no third.) Though it is only the sophomoric type who would bring them up in public, you should be prepared. The following will keep you ahead of the impressionists.

Errors, significance levels, and null hypotheses are all tied together. Let's begin with the last. Contrary to what the name implies, a null hypothesis is not a statement of nothing about nothing, although, very often, it is a conjecture that the truth about something is exactly the opposite of what you wish. In the case of the broken bones and syrup, a good null hypothesis would be "There is no difference in healing times." Thus, researchers are usually excited when the data indicate that the null is false. (Or, to put it another way, it is demonstrated that it is true that what we didn't want is in fact false.)

Now hold on, that's only the beginning. An error of the first kind is what you make if the null is true and you falsely reject it. Furthermore, the, hopefully small, significance level is the chance of making a type 1 error. Naturally, since you hope it is not true, you would not want to falsely reject a null hypothesis that is actually correct.

If your head is swimming, then you're in good company. This

consistent thinking in compound negatives has very likely, over the decades, effected the thought processes of some statisticians. Their published annals tend to be cluttered with this sort of thing. Any wonder that they have trouble convincing others of their wisdom and even argue among themselves? Could it be that the science reporters are really trying to explain what they think they understand from all of these conflicting hypotheses?

One last detail before leaving this mental maze. An error of the second kind, just the opposite of the first kind, is committed if you falsely accept a null that is not true. This might happen if you insist on a very small significance level or the level that is calculated is wrong. (Significance levels are also subject to errors.) This, too, has a chance of occurring, but no special name. As with confidence and accuracy, the chances for these two kinds of errors can be traded. In this context, statisticians refer to the significance level or chance of a type 1 error as alpha (another Greek letter, α), and the chance of a type 2 error as beta (β). Like accuracy and confidence, as discussed earlier, once the data are taken, alpha and beta are locked together. Shrink one and the other will grow. You could, if you wished, pretend that the difference found in bone healing times was a little off, and say "maybe it was only four days instead of a week." If you calculate the new alpha value, it will be smaller than the first one. In other words, it is more believable to say that the difference is not as large as you first thought. This sounds good but, now your chances of a type 2 error change, and not in a very nice manner. For example, in the bone healing experiment, say that both alpha and beta turned out to be 5 percent with the one week difference in healing time. Not being happy with that, you could make the four-day assumption and find perhaps that alpha, the significance level, dropped to 1 percent. That seems encouraging, but you would then find that beta, your chances for the type 2 error, jumped to 82 percent. Hardly a free lunch! That whopping change in one error, from 5 to 82 percent while the other only moved from 5 to 1, is not uncommon. The trade-off between type 1 and type 2 errors is always this way (nonlinear). Falling back on the analogy of the two bungee cords tied together that we used for confidence and accuracy, it's as though one bungee cord were very thin and the other very thick. If you pull on the pair to make a small change in the thick one, it causes a big change in the other. Achieving any kind of balance or equal numbers for them is very difficult. Most researchers simply do not worry about type 2 errors and probably never even compute beta. Who knows what those values may be for all the risk factors floating around out there?

Take a moment to reflect back on this lesson. Your vocabulary is becoming quite impressive—nearly adequate to converse in the big leagues. There is one more area we should touch on before doing anything radical, however.

So far we have considered studies in which only two items were of interest: A versus B in a campaign, long nights and cool days, cough syrup and broken bones, and so on. You might recognize that these are very simple cases, rare when compared to everyday real-life situations in which there are, typically, a host of factors, not just two. Studies of this latter type are called *multidimensional* or *multivariate,* two more terms that are essentially self-explanatory. If, for instance, it had occurred to the Neanderthal witch doctor that the temperature during the long or short night might influence the temperature the following day, he would have had three items to record instead of just two, and his study would then have been multivariate. In private, you might think in terms of a more comfortable, glamorous, and nontechnical word, such as "multifaceted." Whenever you have more than two things to worry about, your study is multisomething.

To deal with multisomething problems, we need to shift gears, or change our mode of thinking. (A welcome relief about now!) Let's return to the witch doctor again, but this time we must credit him with a few more skills than we might willingly attribute to prehistoric man. In this case he has been recording the temperature, e.g., warm or cool, as well as the length of each night. He now has two pieces of data instead of one upon which to base his guess about the coming day. You might agree right away that, because he has more data, his predictions should be better; statisticians, of course, must determine how much better. Consider the following:

Before the caveman started taking any notes at all, he had no data upon which to base his guess about what the next day's weather might bring. (We will assume that he was unable to perceive whether it was winter or summer, or didn't know the difference.) If he had been pressed to foretell the next day's temperature, he would have had to resort to the Neanderthal equivalent of coin tossing. And like any toss of a coin, in the long run he would be right half the time. More to the point for our purposes here, he would be *wrong* half the time (50 percent in error). As we discovered earlier through the archaeologist's diggings, by recording the length of darkness and using whatever his version of correlation was, his accuracy improved from the coin-tossing 50 percent to 85 percent, which means only 15 percent in error. Now we would expect that by adding nighttime temperature to his records, his winning percentage might rise to perhaps 95 or 96. Statisticians, of course, prefer to talk about errors. They focus not on how much better off our caveman might be, but on how much less worse off he is. The emphasis is on error reduction, not accuracy improvement.

The case of the prehistoric weatherman is simplistic because his result is binary; he only has two choices, either warm or cool. His scoring is also binary, either right or wrong. We can still talk about his error rating, but otherwise it's a pretty dull situation—compared to most real problems. Much of the time we really want to predict something difficult, say, what will the actual temperature be, not just whether it will be warm or cool. Then, instead of just shouting right or wrong, we can really feel superior by proclaiming how much: "Ah-ha. It was one degree hotter than you said it would be."

It's likely that the witch doctor began by simply noting that long night equals cool day and so on. Soon he corrected the guess according to whether the night was warm or cool. With a few years of experience, he might have invented several schemes for predicting high temperatures from the kinds of nights. For example, realizing that the longest night is no guarantee of the coldest day, he could moderate his forecasts using the temperatures of a few preceding nights. Eventually he'd evolve to be a true sophisticate, devising formulas and incantations using both lengths and types of many nights.

In many cases, researchers are not that clever. They choose to stick with simple methods, like the witch doctor's first attempt. The formal terms for what they do are *linear* or *straight-line analyses.* These methods assume that all things are *directly related* (they go up and down together), or *inversely related* to one another (one goes up while the other goes down). Sometimes this linear analysis is not bad. More often, though, it's like using a yard stick to measure around the inside of a barrel. See what would happen if the witch doctor used this approach and began his record keeping at say, the beginning of spring. His notes would show how the nights get shorter and the days get warmer. If he drew a picture of this it would be a gently rising line. A nice, consistent, linear behavior. By the summer solstice he should be doing pretty well with his predictions, but then weird things begin to happen. As you know, that's when the nights start getting longer but the days keep getting hotter. The gently rising line, though still rising and getting warmer, now points the other way, toward longer nights. In a reasonable time, our ancient buddy would note this and correct his predictions accordingly. This is where he might start to use some historical (last year's) data to modify his oversimplistic method. Seldom does the researcher or statistician try that sort of thing; he loves straight lines, particularly in multisomething studies. What he winds up doing is using a whole bunch of straight lines, one for length of night, one for type of night, and one each for any other factors he might think of. Of course, as we've learned, he would keep track of his errors to see how well he was doing.

As a rule, errors are a mix of positive and negative values, that is, sometimes the guess is too big and at other times it's too small. That's not an accident; the researcher works hard to invent straight lines that produce an average error of zero. That is, in the long run, his positive value errors will equal his negative value errors giving an average of zero. Until you have gained some experience, a zero average error may sound pretty great, but think about this: two predictions, one with an error of a zillion in one direction and the other with an error of a zillion in the other direction, will, on average, be right on (zero average error)! Remember my friend with his head in the oven. Don't become enamored of zero averages. Our caveman could do that well just by repeating every day: "It's going to be warmer than yesterday."

The better researcher is aware of this so he keeps track of his errors in a special way. He squares them before adding them up. The squaring eliminates minus signs (and zero errors). When you do this, every error increases the sum of the errors (none of them subtracts). For example, suppose you had five measurements with errors of 1, 0, –2, 2, and –1. The sum of those errors is zero, and their average value, the sum divided by 5, is also zero. Just stating that average of zero rather hides what really happened. So, square those errors and get 1, 0, 4, 4, 1. The sum is now 10, which certainly doesn't hide anything. If you think however, that squaring exaggerates the real error size, we can take the square root of 10 to get 3.16 and divide it by 5 for the five measurements, as we did in calculating the average, and get 0.63. This last number, a kind of average-square-root of errors, is calculated exactly the way the standard deviation of your two-wood drives was calculated. But, instead of calling it a standard deviation, we now call it a *standard error.* This is one of the more logical terms in statistics. It is not seen very often because the *sum of the squares* has all the same information and that is sufficient. There is no need to divide and take the square root. So, you may hear the statisticians or researchers discussing sums of squares of errors, or SSs. Then, instead of talking about error reduction, they can consider smaller SS. Now, because the error in any prediction is what is left over after you take out the prediction, the error is called a *residual.* For example: You predict 75 degrees as the high temperature tomorrow, but the high turns out to be 90. Take the 75 out of the 90 and the error, 15, is a leftover or residual. Since even researchers don't like leftovers, the name of the game is "minimize the residuals."

Minimizing residuals with straight-line analyses is one of the shining lights in statistics. We call it *linear regression analysis* or simply *regression.* There are other nonlinear kinds of regression, but they are difficult and often get special names, as we'll see much later, so nobody's confused if you just say regression.

Regression is kind of fun. It's reminiscent of little children drawing trees, in the winter after the leaves have fallen. It is hard to get all those intricate twigs and branches, curls in the trunk, and so on, so what does the child do? One straight vertical line for the trunk, one or two more straight lines angling off for branches and, voilà, a tree in the winter. What's interesting is, almost everyone recognizes it without hesitation. We might judge that one drawing is better than another because of the number of branches or their placement, but they all *look like* trees. The differences are a matter of degree.

Now, when the child draws that first vertical line, we have little idea what is coming. Each new line clarifies the picture, up to a point. Some number of lines is enough and adding more won't really help. This is just what linear regression does. It (usually a computer package) draws straight lines through the clusters of data to show where those clusters lie. Each line is like a branch of the tree, only approximating the twigs (or data points) on it. Each new line comes closer to some of the data than others did, and so the residuals, the distances between the line drawn and where the data (tree twig) really is decreases. So, each new line reduces the SS. This is the kind of thing that makes researchers feel like Nostradamus, the famed sixteenth-century prophet. Keep plugging in more straight lines (more factors to add to the multifactors) and watch that SS head toward zero. At least that's the dream. The computer doing the work doesn't give a hoot that the latest factor may be irrelevant garbage, like my distance from the equator, it will go right on reducing the error, or trying to. Authors seldom justify the lines they use, they just say what they did and then brandish the small SS.

Like the child's drawing when more lines no longer help, regression suffers greatly from the law of diminishing returns. The error reduction gets distressingly minor as more factors are tossed in. There seems to be some significant, irreducible SS. This has been called by some the inherent cussedness of nature. My experience indicates that much of what is written off as intrinsic cussedness is actually more of an expression of the analyst's lack of knowledge, tools, and insight. Mother Nature doesn't talk back, so it's easy to accuse her of excessive fickleness.

As I said, regression can be fun, and we will use it extensively in the next lesson, a real life example of data analysis, including correlation, linear multivariate regression, and some other awesome sounding tools. You will not only see how they are used, but how easily they are misused. You may become a little more critical of the columns reporting that vitamin z reduces the risk of cancer.

Lesson 3
Example

Now that you've read about the rules and a few of the high points of their use, you may be thinking that you know something about statistics. Don't be hasty, it's a big jump from knowing a few definitions to actually applying them. Few people would volunteer to land a space shuttle after reading a manual, yet it seems that many are inclined to use statistics after only a brief introduction.

Seriously though, you will be more comfortable with the concepts and methods of the first two lessons if we plow through the detail of an example. The one I've picked is especially useful because, depending on how you look at various parts of the data, it's possible to "prove" and then "disprove" a number of things. It's the sort of topic that could keep researchers fighting for decades. Actually, the subject, deaths due to car accidents, is so familiar to us that there is little room for discovery. However, this familiarity makes the example all the more cogent. In addition, you have a leg up on this topic because you already have a pretty good feel for the kinds of things that might influence such a grim statistic of modern transportation. You might also enjoy comparing your thoughts on this matter with what the statistics show.

As we indicated earlier, preconceived ideas and prior knowledge, real or fancied, tend to bias your analysis. This could be a real handicap in your examination of the accident data and other related information I'm going to show you. In order to minimize this risk, and to attain some objectivity in this example, try to imagine that you really have no idea of what might affect auto accidents and the resulting deaths. Someone has simply asked you to look at some collected data (five sets in fact) to see what you can make of them. (This will be a descriptive statistical study, as opposed to one in which testing is done.) The person also says that more data will be available soon but you should start to work on these. We can try to simulate ignorance of the subject by hiding most of the real names of the sets of data and just label them "Car Occupant Deaths," for the first set, and A through D for the remaining sets. We do allow knowledge of the fact that the data are ordered by the year for which they were taken. So, we have four unknown items to investigate for their relations, if any, to the first item, Car Occupant Deaths (COD).[1] The years for these data are 1960 through 1974.

Now, light up your meerschaum, pull on the deerstalker, let's play Sherlock.

The first thing any reputable researcher should do, but seldom does, is *look at the data.* This does not mean reading the numbers. What has to be done is to plot the data and look at the pictures, just to get a feel for what is going on. Unhappily, plots throw many folk for a loss. If you're one of those who joyfully skipped both algebra and geometry, viewing the pictures as art may help. However, I know that you at least have some feel for numbers or you would have given up our discussion long ago. Therefore, understanding a visual representation of them cannot be all that difficult.

Since the data are chronological, 1960 to 1974, it is reasonable to line them up in some manner with the years they represent. So, along the bottom of the picture we line up the years, marked with vertical lines. To make all the plots similar, we make a scale on the left edge of the picture with the bottom of it corresponding to the smallest possible, or minimum, value for each of the five data items (the variables), and the top of it corresponding to the largest possible, or maximum, value of each. The values are identified on the plots with the numbers 0 and 1. (Making the minimum equal to 0 and the maximum equal to 1 is referred to as *normalizing the data.*) Each year then receives one point in the picture, somewhere between the top and bottom, according to the actual value of that variable for that year. Connect all the points together with a line and shade in the area under the line. Note that I'm going to cheat a bit here because some of the plots will never make it to 1. (I'm anticipating larger numbers in the next bunch of data.) So we make up the five plots. In this case we'll shade the known data (COD) a little

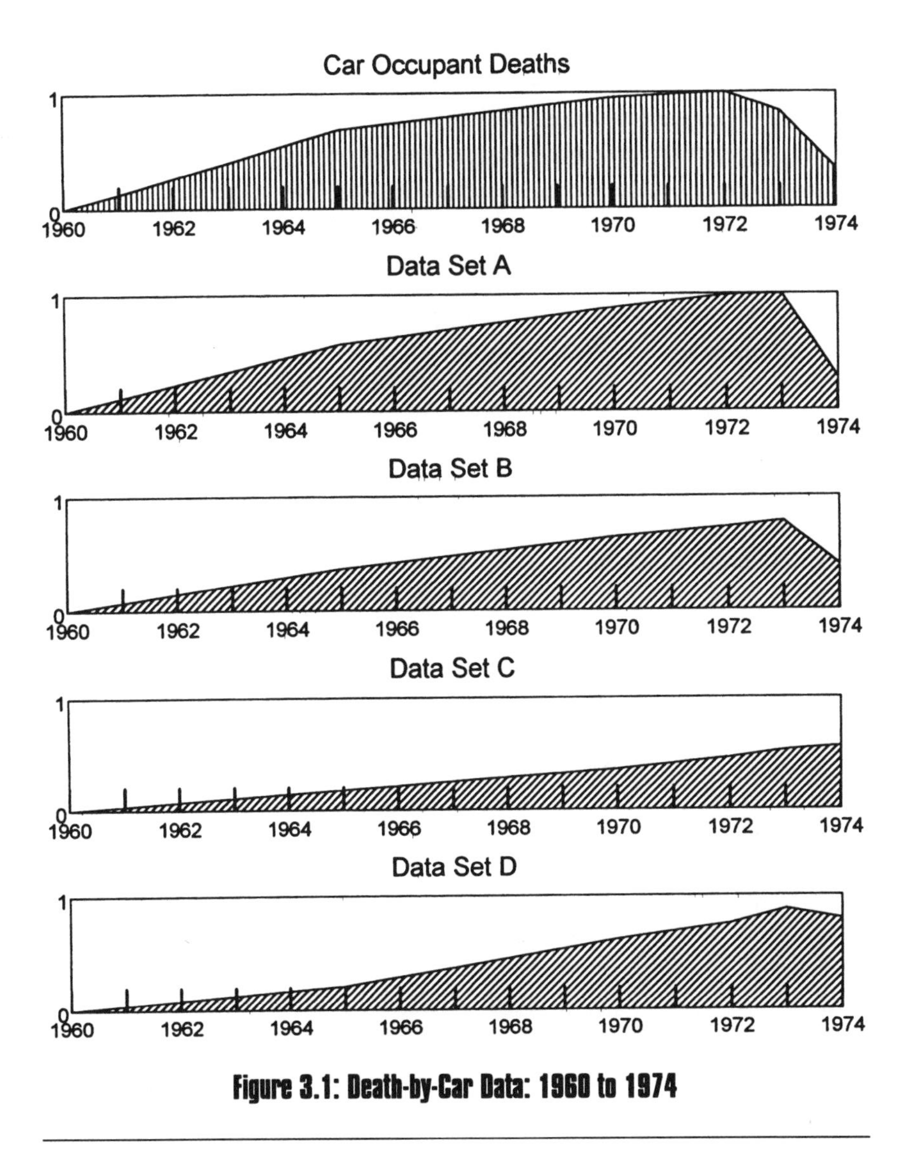

Figure 3.1: Death-by-Car Data: 1960 to 1974

---

differently than the unknowns (A through D), just to keep things in perspective.

Having done all this, we get the very exciting presentation in figure 3.1. Why do I say exciting? Well, my gosh! Figure 3.1 is loaded with positive correlations! Look, all the plots start out small in 1960 and grow, in concert, continuously until the early 1970s. Then most of them fall off together. Something significant is going on here, even Dr. Watson would say so.

Because of this similarity of all the pictures, at first glance it appears that any or all of the four mysterious variables, A through D, may be closely related to death by automobile.

Let's do some correlations, you say? Hold on, don't go running off half-cocked. You're beginning to act like a media-motivated researcher already, and you don't even have your feet wet. Relax and look at the pictures some more. There may be other items to note, clues to direct our upcoming analyses.

Now the first plot shows how the COD varied from year to year from 1960 through 1974. Note that the number of deaths just increased steadily to a peak at 1972. This peak in car accident deaths in 1972 is a bit of fairly well-known trivia which might be used in invoking rule 1 (extol the obvious) when a report is written. Several plots other than COD also have peaks. However, none has its peak in the same year as the maximum number of deaths, 1972. They all come later. That is a point to investigate. In addition, we might ask, why do most of the curves rise until about 1972 or 1973 and then fall off? A and B go into real nose-dives. Did that drop in deaths after 1972 trigger some actions that caused the other, still unknown things, to fall off? These are only a few of the many questions that will pop up and beg to be explained.

Besides attempting to answer these questions and others, we should, like good researchers, strive to find a way to predict the number of deaths if given some of the unknowns or, in the greatest of all scenarios, discover how to predict them in years to come based only on history. By the way, predictions of that sort are called *extrapolations*. For example, your experience or history in buying cars tells you that prices generally increase from year to year. You may want a new car but wonder whether you should buy now or wait for the new model next year. Down inside, you know that if you wait, the car will probably cost you more, and in some manner you weigh the benefit of the new model against the expected increase in price. That "expected" increase in price is your personal extrapolation of car prices, based on your own history. Everyone knows that extrapolation is dangerous, that new model car might actually cost you a lot more than you had expected. Nevertheless, the urge to know the future is overpowering so we extrapolate all the time. Very seldom is anyone seriously taken to task over a wrong extrapolation. Excuses for missed predictions are available by the barrelful. Just listen to the economists explaining why the recession did not end on time or inflation last month went down instead of up. One of the easiest ways to make predictions is to use the linear regression techniques we talked about at the end of the last lesson. We shall do that, but first, now that we've had a closer look at how the data are behaving, let's do those correlations.

Mechanically, this is duck soup. Dump all the data into the computer and out will come a whole table of correlation coefficients (those numbers between plus and minus one), relating every variable to every other one. It doesn't matter if they are sea water and table tops, or coal mines and fish eyes. As long as any numbers relating to coal mines (for example, how many there are in each of the fifty states) are put in one column and numbers relating to fish eyes (say, measurements of their size in fifty species) are placed in the next column, the computer will generate a correlation—number crunching has no conscience. Computers have done us a real disservice here. Before computers (BC), analysts spent a great deal of time poring over the data, examining detail, looking for apparent relations and so forth, before expending large amounts of time and patience with pencil and paper or mechanical calculators to derive correlation coefficients or other statistics. One might say that giving computers to analysts was akin to giving aerosol paints to graffiti artists. Nonetheless, it's hard to resist. For our little study here, the computer supplies us with the following:

	COD	A	B	C	D
COD	1.00	.98	.93	.65	.69
A		1.00	.97	.71	.76
B			1.00	.85	.90
C				1.00	.99
D					1.00

Table 3.1: Correlation Table for Death-by-Car Data

Now don't get rattled by all these numbers, there are many variables, a real multisomething study. Don't fall into the trap of thinking that everything is going to be simple. Learning to live with a lot of numbers is not too tough, I'll take you by the hand through this one and you'll see that it's not so bad. Plenty of variables with lots of numbers can be beneficial in a defense of your work if the need should arise. Like the boxer who dazzles with fancy footwork, you can wear people down with numbers.

Let's look at the table. Notice that the columns and rows have the same headings, the names of our datasets or variables. Each number is the correlation coefficient between the variable of its column and that of its row. The first number 1.00, for instance, indicates that column COD and row COD have a coefficient of 1.00. As we noted earlier, that's as it should be for any variable correlated with itself. The number 1.00 indicates ideal or perfect correlation. See all the 1.00's in the diagonal?

Now look at the first row, COD, and the last column, D. At their intersection we see the number .69, the coefficient between our car occupant deaths and the unknown D factor. The number .69 happens to be a rather "good" value, a fairly high correlation in statistical studies. For now, don't be concerned with the meaning of the number, that is subjective and you will develop a "feel" for that as you encounter the many examples in the rest of this book. With a little practice, you can scan those rows and columns without batting an eye and read off these coefficients. If you have the time, memorize some tables like this, they may enhance your cocktail party risk factor tidbits.

Recall that the coefficients can take on values anywhere between plus and minus one, and that zero is (almost always) a perfect bust (in other words, one variable doesn't relate to the other at all). But now, look at all those really big numbers in our table: .97, .98, and .99. They verify what our look at the pictures told us. And, yes, they are about what you thought they would be when you first asked for them. There are strong ties between deaths by car and whatever A through D are. A correlation table with numbers like this is a researcher's dream come true. In fact, it is so good that experienced professionals would be very skeptical: "Something is fishy here."

Now let's see what items (variables) A through D actually are. The first row in the table gives the coefficients between deaths (COD) and each of the other variables. I'll tell you what they are in that order. A, with its coefficient of .98, and the plot right below COD in the figure, is the average speed of all traffic during the year indicated.[2] Look at that plot of A again. Notice that from 1960 to 1973, we drove faster and faster. The actual numbers, according to the federal bureau that keeps track of these things, go from 54.8 to 65 miles per hour. Why did this happen? Your guess is as good as mine. The interstate program was completed at about that time—maybe that had something to do with it. But such statements are pure conjecture, worthy of little more than a suggestion to make another study to investigate that possibility. So, A is speed and it has the highest correlation with deaths. Ah-ha you say, I knew it, speed on the highway is a killer. Hurrah for President Richard Nixon's 55 mph speed limit!

If the study were to end here, you could use it to drum up support for more highway police radar or more officers to patrol. Unwittingly, besides equating correlation to cause, you have just fallen into one of the first and most common traps set for the researcher. You found exactly what you expected to and your blinders are now nearly cemented in place. (You won't really appreciate that statement until near the end of this lesson.) Let's proceed.

The next plot in the figure, variable B, relates to COD with a coefficient of .93 according to the table. That's a whopping large number also, so it must be important in the death by car phenomenon. As it turns out, B is the total number of car accidents in a given year.[3] Do I hear a disappointed "Oh"? With more accidents occurring there would probably be more deaths. No real surprise here. Maybe so, but take a close look at the pictures. If more accidents (and more speed) automatically equate to more deaths, why do both speed and accidents increase from 1972 to 1973 while the number of deaths decline? There I go asking embarrassing questions instead of moving along with the correlations. In cases like this, it could happen that you never discover the reason for the event. If you do not care to admit your ignorance, just invoke rule 1, part 2: Ignore the obvious.

The next dataset on our list, C, is different. Note that it does not drop off near 1973 as COD, A, and B did. So what is C? A reflection of the increasing population and standard of living, C is the total number of cars registered for use, and it just continues to climb.[4] Now, if you think about that for a minute, especially in the time from 1972 to 1973 when the number of deaths declined, it just compounds the question I raised. That contradictory trend—the decline in deaths after 1972 while the number of cars on the road was still climbing—is reflected in the relatively low (for these data) coefficient of .65. The number of cars and the number of deaths are going in opposite directions in those last years. If that kept up long enough, say for another ten years, the coefficient would go negative. Yes, when the coefficient is negative, it means that the two things are closely related but move in opposite directions, like our Neanderthal's increasing darkness but decreasing temperature as winter approached.

Does this appear to be going nowhere? After the revelation that speed and death go hand-in-hand, things have been pretty dull, in spite of the apparent contradictions noted. But let's finish this off. D is simply the total gallons of gas consumed annually.[5] It appears from the table that the total gas usage tracks closely with accidents, B, and of course, total cars, C.

Remember that I said all the plots were normalized. That means that the total variation in the thing plotted was spread over the whole picture, top to bottom. When this is done, it looks as though everything starts at zero and that all the changes in the items plotted are in some sense equal. That is, they all go all the way from the bottom to the top (or nearly so in this case because of my cheating), so they must be equal, isn't that so? Normalizing is handy and useful when it's only the relative changes in things that are interesting (car deaths related to increasing speed levels). On the other hand, such normalizing can be very deceiving because it throws everything out of context. If we look back at the real numbers, which I did not show you, the true changes from the minimum to maximum values in those plots vary from 15 percent for deaths to 90 percent for the gas used. The changes are actually very unequal. And note that although the item of major interest here, deaths, changed the very least, the normalizing made it look gross. Normalizing is handy to make inconsequential changes look really drastic in a plot as the COD data appeared here. The media folks love to use that, and similar techniques, to draw impressive looking charts presenting insignificant data. Be extremely cautious when you see a plot in the newspaper showing very bad or very good changes in something. See if it is normalized, then try to find the *real* changes.

Let's now redraw all of the pictures with the proper names of the datasets inserted in figure 3.2 so that you can digest all this a little more easily. Some of the items mentioned above are also noted in the figure.

For the moment at least, the correlation business seems to have been run into the ground, so let's turn to regressions and predictions (extrapolations). That has to be more exciting. When we first mentioned regression in the previous lesson, I hope you didn't go to *Webster's Dictionary* to look up the meaning, which only provides the better known definition: a backwards progression or retrogression, physically or mentally. The use of the term in statistics derives from that meaning. Some one hundred years ago, a statistician used the word to describe a negative characteristic of human inheritance

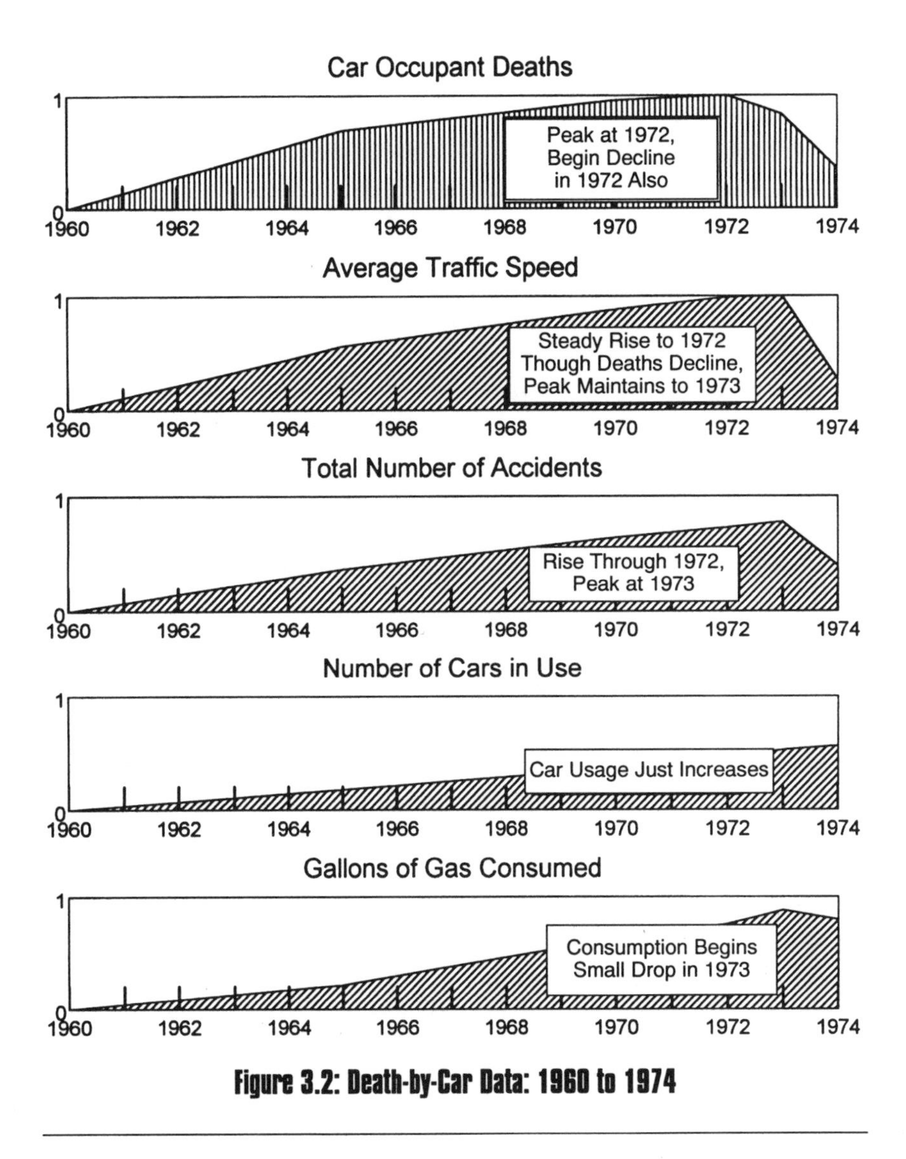

Figure 3.2: Death-by-Car Data: 1960 to 1974

---

that he had pinpointed through a linear analysis. Others in the profession apparently liked its sound and imposed the current statistical usage with its (forward) predictive connotation, apparently disregarding its retrogressive roots. So, in a sense, we wind up making predictions by looking backwards. We'll see whether this brand of hindsight is 20/20.

These death-by-car data provide many opportunities to illustrate the strengths and weaknesses of regression analyses, even multisomething regression analyses. Actually, we have no business attempting any multivariate regression with these data. They violate an underlying premise that all variables or factors must be independent of each other. That technical phrase, "independent of each other," is jargon which means roughly that, in our case for example, if more miles are driven, the amount of gas used need not change. Obviously, for most of our variables A through D, it is difficult to see how changing any one could not affect some or all of the others. When the factors are not independent (when you can't change one without affecting others) the whole theory behind multivariate regression is undermined.

To understand this, consider the trivial case of relating wear on your auto's tires to other variables. To keep it very simple, let's assume that you only drive on interstate highways, exactly at the speed limit. It should be obvious that you could do a regression analysis of tire wear against miles driven, or that you could do one of tire wear against gas consumed. If however, you tried to do a multivariate regression of tire wear against both miles driven and gas consumed then, under ideal conditions, the computer computation would fail. You would get a message to the effect that the calculation is impossible. Under near, but not ideal conditions, the computer would give an answer but it would be totally meaningless. Let me explain.

By ideal conditions I mean that your car is extremely consistent in its performance and, let's say it runs on a kind of auto-pilot. Thus, the miles per gallon value will never change or vary. If you tell me the miles, I will know exactly how much gas you needed and vice-versa. The two sets of data you give the computer, gas and mileage, are completely dependent, one on the other, and redundant. All the machine needs is one. The mechanics of the arithmetic in the computer cannot handle the duplicated data and will break down. In a more realistic case, the miles per gallon for your car will vary a little. The data will not be completely redundant, only nearly so, and some meaningless answer will result. Depending upon how closely the gas and the mileage track one another, you may or may not be able to tell that the results are meaningless. The COD example of this lesson is a case in point. The numbers generated here appear to be "good."

This technical point about independence is, by the way, often the most difficult thing to check or verify in real life research. It frequently becomes necessary to take the same approach as we did for unknown distributions in lesson 2 where we assumed that anything unknown would be the most common one—the normal. Instead of saying, "Assuming normality," the researchers glibly state, "With independence of these factors." In other words, we'll assume that they are independent, and because proof, one way or the other, is generally impossible, not a great deal of controversy results. Any attacker of the assumption is in an inferior position, no matter which way the argument is going. The defendant is akin to the legal one, assumed innocent (correct) until proven guilty (wrong). In the rare case where you can prove that the assumption of independence was wrong, all regression studies and much of the other analyses will be instantly rendered null and void. This simple fact seems to be consistently ignored in a great many statistical studies. In the hundreds of statistical medical study reports I have examined, only a few have even mentioned the word *dependent,* and only a small percentage of them mention testing for *confounders,* a procedure which helps to uncover dependent variables.

The sad thing is that if two variables in a study are dependent, as parental medical history and high blood pressure might be in a study of heart disease, any risk factor based on a multivariate study including those two would be wrong to some extent, just as our estimate of tire wear based on both gas consumption and miles driven was. This very point was overlooked in an examination, by a statistical consultant, of two conflicting studies of heart disease that I will discuss more a little later in this lesson, and again in lesson 6.

By the way, it is common to see the term "correlation" used when dependent is meant or vice-versa. The two are not the same. Independent things are often correlated, e.g., snow and long nights. These share at least one common cause but are not dependent; for example, as the latitude varies, there may or may not be snow even when the nights are longer. The presence of snow is dependent upon the latitude. Conversely, very dependent things can have a zero correlation. A silly example is presented by railroad tracks. Consider the positions of the left and right rails as the track wanders around the countryside. Well, they are always the same, one is 28.25 inches on one side of the center line (in the United States), the other is 28.25 inches on the other side of the center line. The position of the rails is completely dependent on the center line of the track. But, if you calculate the correlation between the center line and the two rails it will be exactly zero.

Pondering these last few remarks may give you some idea of why it is often impossible to find out if variables in a study are dependent or independent. The use of these words is so sloppy that theoretical statisticians have published papers talking about dependencies, even in the title, when they really mean correlations.[6]

The game of blackjack provides good examples of dependent and independent events. Card counters rely on the dependent relations that arise. If you are being dealt the first card from a fresh, well-shuffled deck, the chances of being dealt an ace are four in fifty-two. The same would be true for a king or a queen, of course. Say that you did in fact receive a king. The chance of a king on the next draw is now only three in fifty-one because the deck has one less card and one less king. The chance of the second king is dependent on the first king. The chance now for the queen is still four, but out of fifty-one. If any card except a queen was drawn first, the chance for the queen on the second is still four in fifty-one. The chances for the queen on the second draw are independent of whether a king was drawn first, and only dependent on the fact that a queen was not drawn first. This is what the card counting is all about, keeping track of how many of each type card has been dealt so you know how the chances for each are changing. This reasoning leads to one formal definition of (statistical) *dependence*. When two events are considered, they are dependent only if the chance of one of them changes with the occurrence or nonoccurrence of the other. Card games are loaded with dependencies.

In any case, even though the following is statistically illegitimate, it is illustrative and highly instructive of the ins and outs of regression.

Consider the variable we called A, average traffic speed. From the pictures we drew, it is apparent that deaths and speed track one another closely and we saw that the correlation is a very strong one (.98). Recall that the statistician, who likes to be quantitative, measures his uncertainty with SS (sums of squares of variability, or errors) and uses reduction of SS as a criterion of success. Recall that smaller SS means less total error. You ought to develop a warm feeling about this SS business, and now is a good time to do it.

If we use average speed to predict deaths via the regression route, we need a starter SS, some number to measure how good the predictions are if we come up with something that gives us a smaller SS. Now, if the COD never changed, there would be no variation, and the first plot in figure 3.1 or 3.2 would be a simple horizontal line from 1960 to 1974. Suppose now that we take the average value of COD for those years and use it to plot just such a line, superimposed on the actual plot of COD. The differences, or spaces, between the straight line and the real plot show how much the COD varied compared to its average value. This difference can be thought of as the total or intrinsic variation in COD. We can also think of that variation or departure of the COD from the average as an "error," as it would be if we just used the average to predict the deaths for all those years. If we then take the errors for each year, square them and add them up, we have a starting value for SS. It is a measure of the total departure of the actual plot from its average value. In this sense, the starting SS is a measure of all the variation in the COD. This concept is pictured in figure 3.3. The COD has been normalized to show the average value as zero and so emphasize the fact that there are both positive and negative "errors" around that average. The thin vertical lines, one for each year, represent the annual departures from the average that we square and add up to get the SS. You can also think of the SS as being (the square of) all the space or area between the two lines. You can see that, if the SS (that area) were somehow reduced to zero, we would have achieved perfect prediction. The goal of regression analysis is to find better ways than the average value to predict the COD. We can measure how much better each way is by comparing the SS for each new prediction with that represented in figure 3.3.

For the data plotted in figure 3.3, the SS turns out to be 1.476. This number, by itself, has absolutely no meaning. It is an arbitrary benchmark, nothing more, but is useful as a reference for the quality of our regressions. I will refer to that value, 1.476, as the intrinsic or native variation of the COD data.

The regression analysis is done in the same way that we did the correlation: Pump the data into the computer. In this case we get back all the pieces of an actual formula that yields a guess of some number of deaths for any speed we supply to it. Other numbers may also pour out of the machine when we ask for a regression. A popular one among the analysts is something called the *coefficient of determination,* which has to do with some relative internal properties of all the numbers, not how determined an analyst someone is. Textbooks almost always just name it, tell us how to calculate it, and say something to the effect that it's nice if that number is close to one (as opposed to zero). For our little example here it comes up as 0.953. And, of course, we'll get back to that shortly.

The formula that the computer came up with should predict deaths if we supply the speed. If that is done for the speed in each year, the result is shown by the striped line in figure 3.4. The origi-

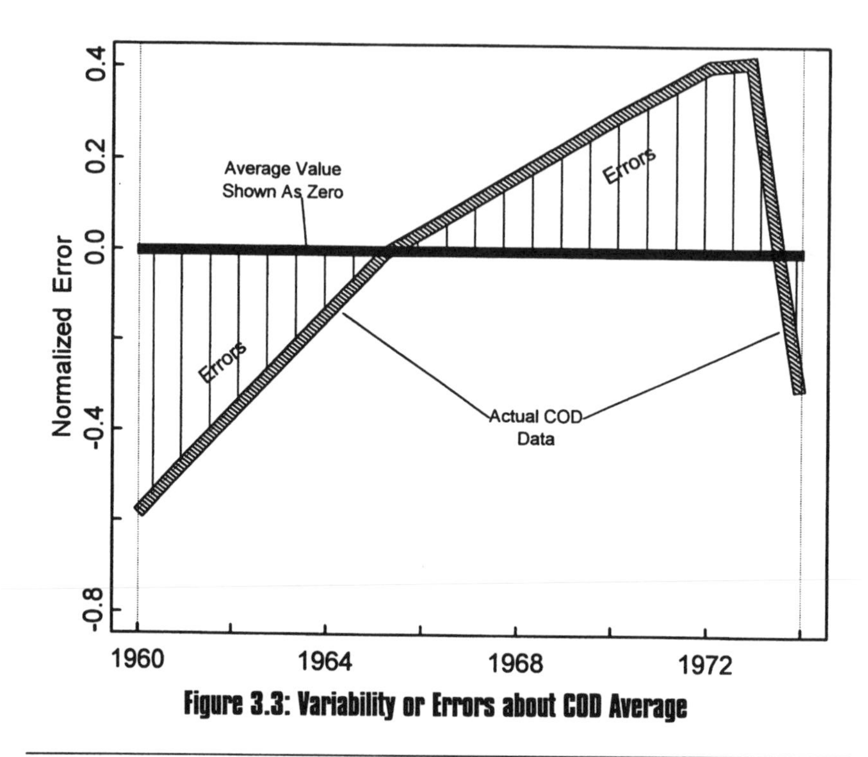

Figure 3.3: Variability or Errors about COD Average

---

nal COD curve is shown with the solid line, and the errors are indicated once again by the thin vertical lines. In this case, the SS corresponding to the error area is 0.069. Subtracting that from our earlier total native variation of 1.476 indicates a "reduction in uncertainty" of 1.407 which is 95 percent of the 1.476 total. That means, approximately, that the prediction line in figure 3.4 is about 95 percent better than the solid (average) line in figure 3.3 for guessing at the number of deaths. That's a significant improvement. That relatively small number, 0.069, describing the area of errors in figure 3.4 simply confirms what we thought we already knew by looking at the pictures: Speed is a reasonable predictor of death. However, now we can be quantitative about it; we have a formula and real numbers to back up what seems to be obvious anyway.

Of course the statistician refers to that 95 percent as a reduction in uncertainty rather than an improved accuracy. Now note that if the number (it is 95.3 rather than 95 if you use one more decimal place) is divided by 100, it is the same (0.953) as the coefficient of determination we mentioned above; so that's what that fancy name

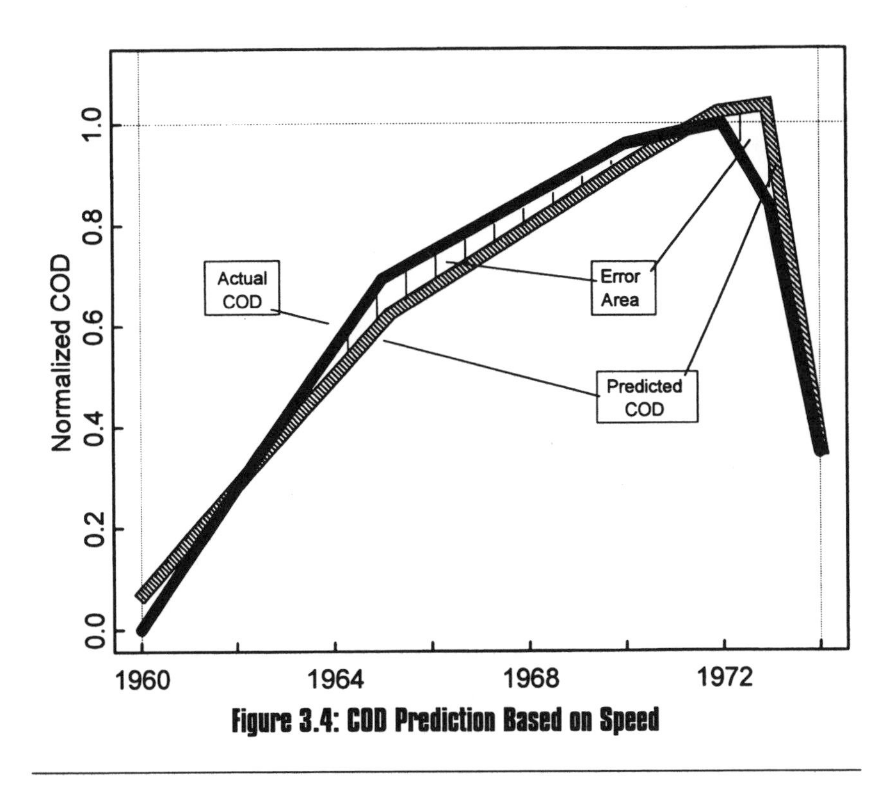

Figure 3.4: COD Prediction Based on Speed

means. The *coefficient of determination* is the reduction in variability or uncertainty due to the regression. Think of it as showing how much better a prophet you become by doing the regression as compared to just knowing the average number of deaths.

It is informative to pause a moment here for a small aside. The formula that we asked the computer to generate gave us a prediction of deaths if we put in speed. We did that and drew figure 3.4. As we stated earlier, one might use that figure, and the formula, to argue for more speed control. "After all," we might say, "here is a mathematical formula that proves that increasing speed increases deaths." Well, I told you the computer has no conscience and it will just as easily give a formula in which we plug in deaths and predict speed—the inverse of the formula we already have. Would you then argue for "death control" as a way to reduce speed? Real statisticians know better than to talk about cause and effect.

Another mystic number that we have been tossing around, the *correlation coefficient,* can also be uncloaked simply by squaring it. For COD and speed, we discovered earlier that the correlation was 0.98

(extended to three digits it is actually 0.976). The square of that is 0.953, which is, again, the coefficient of determination, or the fractional reduction in uncertainty achieved by the regression. Coefficient of determination, correlation coefficient—remember the names, but don't be confused. Measured reduction in uncertainty is all they're talking about and now it's obvious why it's nice to get the number close to 1.0 or 100 percent; that would mean perfect prediction. If you ever do reach 100 percent, however, you will have somehow transgressed from the world of statistics. Perfect certainty can never be achieved.

Thinking or talking about smaller sums of squares as reduction in uncertainty is actually a dangerous thing to do. For example, in the preceding discussion, the implication was that there was too much uncertainty in the death by car data if we only knew the average, and we removed some of that by looking at speed. That's sheer folly, of course, for at least three reasons: First of all, the number of deaths predicted from the speed data were not exact, as figure 3.4 shows. Nevertheless, figures like that are often used to "prove" that, in this case, speed is a strong risk factor for death. The troublesome thing about that statement is that the term "risk factor" implies prediction, an extrapolation of what seems to have happened in the past. I think we already agree that extrapolation is dangerous, but shortly I'll show you some really bad ones. Second, why try to predict COD from speed if we already know it? Is the speed data somehow better? Finally, the "large uncertainty" of the death data around its average is a completely arbitrary measure, nothing more than an artifact used for comparisons that may or may not be useful. Don't be mesmerized by the casual use of words or names like this. For SS reduction, some people prefer to say that some amount of the variability can be attributed to the other factor, whichever one was used in the regression. The (probably wrong) implication in this case would be that if speed never changed, the number of deaths would be 95 percent less variable. A statement such as this stands a better chance of being truthful than a statement about risk factor that implies more speed equals more deaths. Of course, in a more practical situation, the "death" data to be expected from a newly planned highway would be unavailable and projected speed limits used to make crude estimates of them. In that sort of situation, those coefficients take on more meaning. They might even be useful for studying history, but please don't try to make serious predictions such as what will happen if the speed limits are changed.

Safety-conscious extremists love to do that. A little later we will see how well it works.

Not unexpectedly and even if not exact, we did pretty well in estimating the history of deaths on the highway with that single regression on speed. Figure 3.4 is ample proof of that, but remember that there is no extrapolation here. To continue the example and illustrate some other facets of regression, we continue by first including the next variable in line, total car accidents. This time the computer comes up with a formula that lets us plug in the number of accidents in a year, as well as the average speed. The mystic coefficient of determination turns out to be 0.958 and, of course, the reduction in uncertainty is just 100 times that or 95.8 percent. On a roll here, we throw in number of cars in use and then gasoline consumed and let the computer do its thing. The corresponding reductions zoom up to 96.7 percent and then *all the way* to 97.5 percent, illustrating what we claimed earlier, that multisomething regression rapidly suffers from diminishing returns. With each added variable only a very small improvement in reduction of error occurs. This is also apparent by comparing the error area in the resulting plot, figure 3.5, with that in figure 3.4. The improvement is hard to see. The gain by adding three more variables to the first one of speed was hardly worth the effort, but isn't that what computers are for, to crank out that last little trifle?

Having done so well with these estimates of historical events, we might be tempted, even though it's dangerous, to peer into the future, or extrapolate. The usual way to extrapolate is simply to extend the shaded lines of figures 3.4 or 3.5 in the same direction they were going in 1974, when time ran out in those pictures. We will do it a little differently here, but first let's find something with which to compare our anticipated results.

Recall that whoever gave us these data said more would be coming and they just arrived. We have been given the same types of data as before except that this time they are for the years 1974 through 1986. One thing to check is how well our magic regression formulas, generated from the first bunch of data, will work on these. Before doing that however, let's peek at the new data themselves. Things have changed.

Using the new numbers to make up the pictures in the same way as before, we get figure 3.6, in which the old pictures have been repeated with the new data appended for ease of comparison. Note that the data here pick up exactly where they left off previously, the

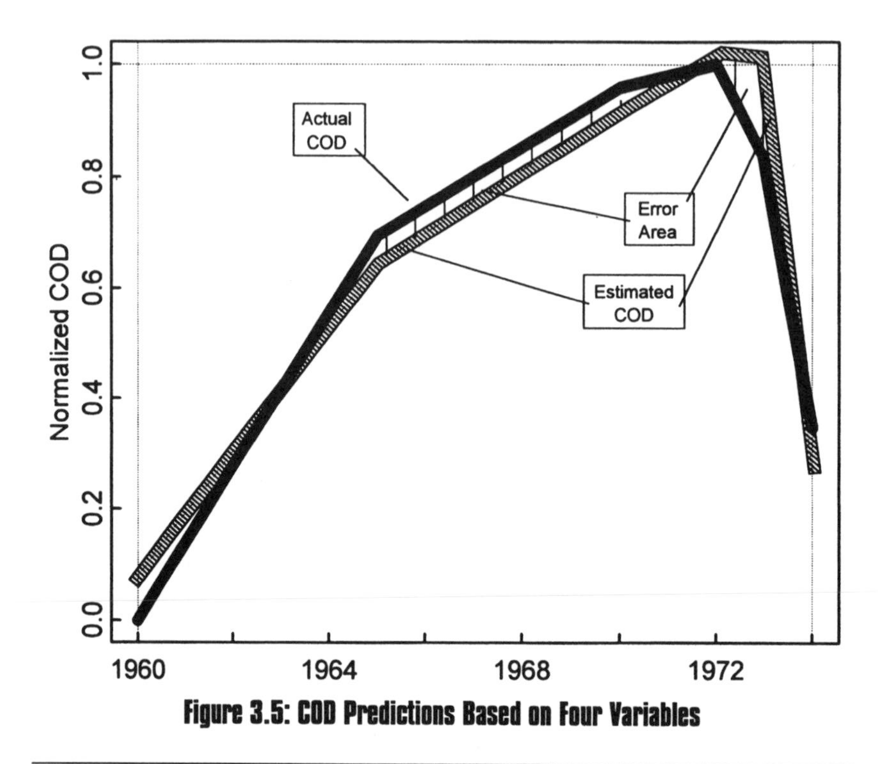

Figure 3.5: COD Predictions Based on Four Variables

year 1974, marked with the dark center line. It's instantly obvious that something happened around that year. The number of deaths leveled off and remained fairly low compared to those from 1960 to 1974. Average speed is considerably slower than in the late 1960s and early 1970s and—here goes your theory about speed equals death—from about 1980 on, speed increased slightly but the number of deaths dropped a lot! Not only that, but the number of accidents even increased during this time. What's going on here? More accidents, at higher speeds, but fewer deaths! Since it now appears that an increase in speed indicates a decrease in deaths, perhaps we should update Mr. Nixon's law to, say, 155 mph. Maybe that would reduce the number of deaths to zero.

Glancing down the rest of the pictures showing the new data, we note that the number of cars has continued to increase but that gas consumption actually dropped after its peak in 1978. You can mull all of this over and speculate to your heart's content about small cars, efficient engines, seat belts, or whatever. It is unlikely that any of *these* data will supply an answer. Besides, I already told you that

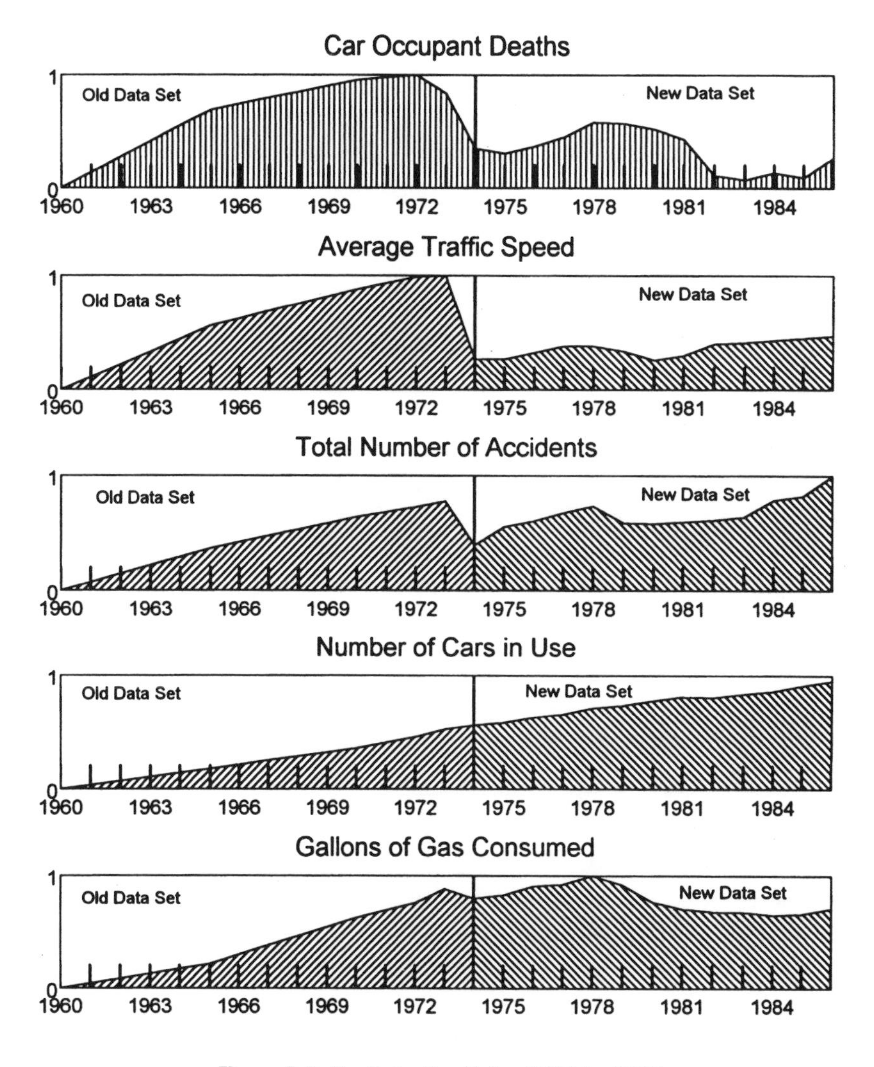

Figure 3.6: Death-by-Car Data: 1960 to 1974

---

speculation is inappropriate in a report, except as a stimulus for future work. On the other hand, you might question, or at least wonder about, the integrity of these new data, or the previous batch if you prefer. At least you had better expand the scope of your report and re-examine those obvious earlier conclusions. If you don't, chances are good that someone else will.

It is time to proceed with the regression lesson. As we said, we

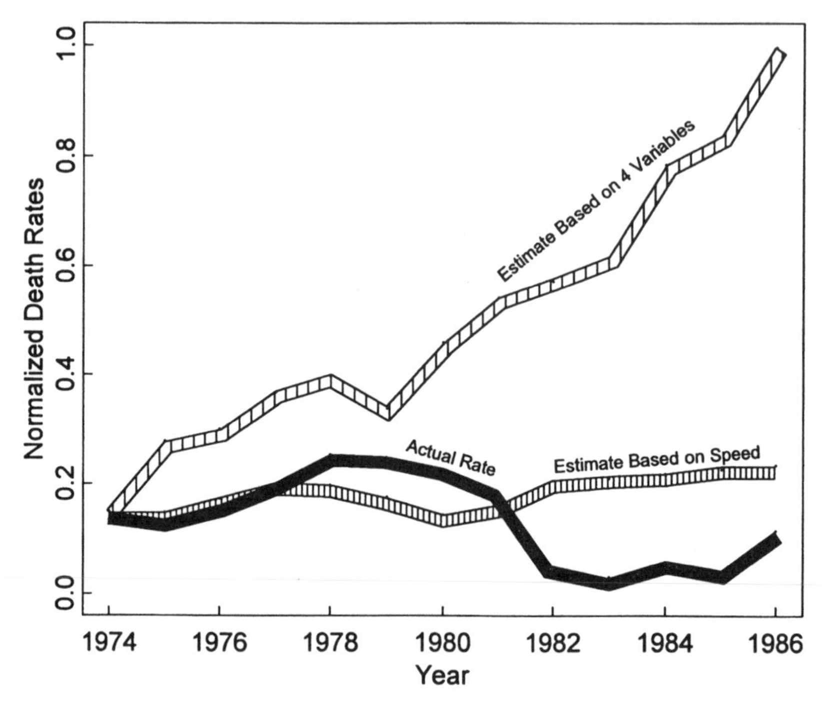

Figure 3.7: Estimates Based on 1960–1974 Data

---

can apply the formulas developed from the first set of data to this new set, just to see what happens. If our magnificent discoveries of strong correlations with amazing predictions of car occupant deaths are valid, we ought to have something here. However, in view of the dramatic changes we just saw, don't hold out much hope for agreement between predictions and reality. If we just plow ahead anyway, with those formulas developed earlier, the results are at least amusing. Figure 3.7 shows what happens when we predict car occupant deaths from 1974 to 1986 using just the average speed formula and then the one incorporating all four variables. The speed prediction doesn't look too bad, but it actually misses by a factor of about two at 1978 and 1983. And the earlier "better" predictor, all four variables, goes bananas. What more proof do you need that extrapolation is dangerous, especially statistical extrapolation? Now in real life, all the advice concerning the use of XYZ to protect against disease is based on nothing but extrapolations of fancy regressions. Are you beginning to see why they conflict? By the way, the computer provides us with the sums of squares for these last two cases, even though they are useless. It says the improvement is –60 percent and –175 percent. The negative signs are of course the clues. They mean that the predictions are 60 percent and 175 percent worse than all the uncertainty in the original data! They are telling us that what we are doing is wrong. To make things even more treacherous, the fancier regression techniques used in many studies do not even provide such obvious warnings. In fact, considerable work is required to ferret out the equivalent information.

Don't be completely dismayed just yet. We can still do a fresh correlation and regression study on all the new data; there could be something to learn here. Plugging the new numbers into the machine for correlations, we get:

	COD	Average Speed A	Car Accidents B	Number of Cars C	Gallons of Gas D
COD	1.0	–.56	–.28	–.46	.80
Average Speed A		1.00	.84	.73	–.35
Car Accidents B			1.00	.78	–.26
Number of Cars C				1.00	–.67
Gallons of Gas D					1.00

Table 3.2: Correlation Table for New Death-by-Car Data

The first obvious difference here, compared to the earlier table, is the presence of minus signs. These represent major changes. For example, it was clear before that speed could be equated with rising numbers of deaths, but now we see that they have a negative correlation. We can reach a new conclusion: Faster is safer! I told you these data were a lot of fun, that you could first prove and then disprove things.

More exciting conclusions can be determined from the table: Greater speed means more accidents (+.84 in row A, column B) but more accidents also equals fewer deaths (–.28 in row COD, column B). In fact, the only thing that correlates positively with deaths is a greater use of gas, and, in spite of our logic, increasing the number of cars makes for less gas consumption. The computer said it, so this must be correct. We may find more interesting conflicts between what this table appears to say and either our perception of reality or the numbers in the first table. Can we explain some of the apparent anomalies just mentioned?

Just picture yourself as the chief of staff of a group studying problems like these. You happen to receive the 1960 to 1974 data, while your colleague and adversary from Contempt U. happens to wind up with the 1974 to 1986 data. Would the journals and general media ever have a field day with the ensuing reports! A real-life case of this type, one of major interest, appeared in the *New England Journal of Medicine* in 1985. Two studies regarding the relationship between the use of estrogen and the incidence of heart disease published in the same issue were in direct disagreement.[7] One claimed estrogen use was dangerous to the heart, the other said it was beneficial. The journal's statistical consultant endeavored to calm the waters with a companion article.[8] After offering several excuses, none of which he chose to believe, he resorted to, in his words, "the investigator's great cop-out: More research is needed." Well, now, ten years later, supported by more conflicting studies, the pros and cons of estrogen therapy for women are still being argued. In lesson 6 I offer some facts and thoughts about why those two reports disagreed.

To make some comparisons with the terrible extrapolations we made, suppose now that we use the new data on speed and so forth to estimate deaths within this new time frame. Since the new correlations are not nearly as strong as the old ones, we should not expect the predictions to be as good as they were for the first case. And in fact, they are not. Using the speed-death formula derived from the new 1974 to 1986 data, we arrive at figure 3.8. These estimates are almost as bad as those extrapolated using the previous formulas; the areas between the curves are of similar size. Note that as bad as these appear to be, they are based on a correlation coefficient, about 0.5, that is rather large by common standards. Many workers are delighted with numbers like that. This is a rather good illustration of the kinds of uncertainties cloaked by common phrases such as: "A strong relation was found between the number of goodies eaten and the length of the left little toe." Whenever possible always ask, "How strong?" and keep Figure 3.8 in mind. I encourage you to develop a strong disregard for "strong correlations."

Before leaving the subject of deaths of car occupants, I must point out that the earlier game of forward prediction, 1960s data-derived formulas applied to 1980s data, can be done in reverse just as easily—namely, backwards prediction. That is, we can take the 1974 to 1986 data, derive new formulas, and apply them to the earlier dataset—a reverse extrapolation. This is often done when attempting to fill in missing historical data. Using the new

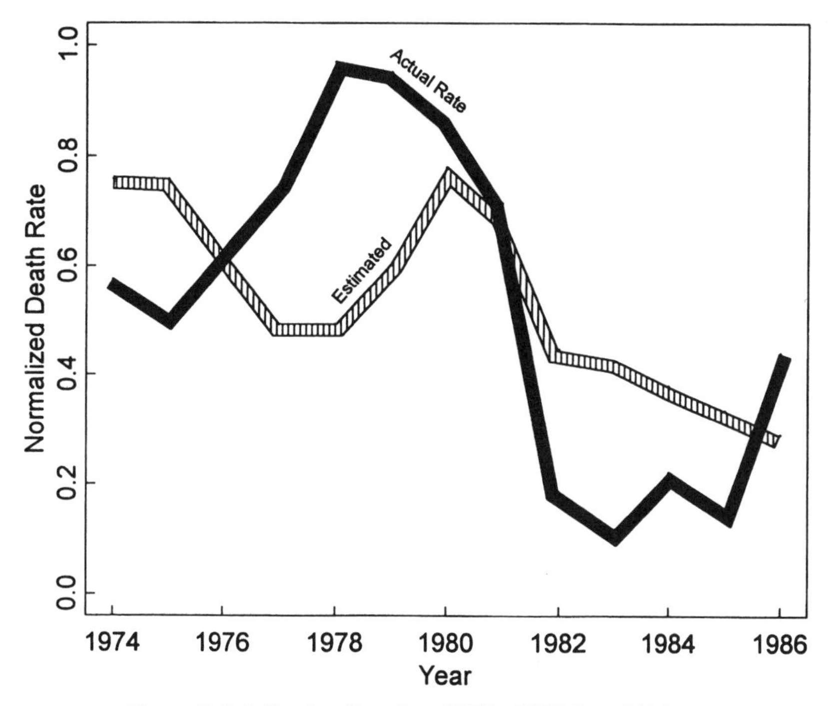

Figure 3.8: Estimates Based on 1974–1986 Speed Data

1974–1986 data to project backwards, in a "what might have been" scenario for the 1960 to 1974 period, we get figure 3.9. The backwards prediction is practically a complete inversion of what really happened. In fact, the formulas predict negative deaths from about 1965 to 1973. Do we have a new medical discovery here? This shouldn't be surprising given that the correlations were positive in the original data and are now negative. The change from positive to negative comes about from the fact that, for reasons unknown here, deaths in car accidents decreased in the new time period even though accidents and other factors increased. We now expect one variable to decrease when the other increases.

As a last example of how correlations (and thereby researcher's conclusions) can vary, the correlation coefficient between car occupant deaths and average speed is calculated for different years between 1960 and 1986 with these results:

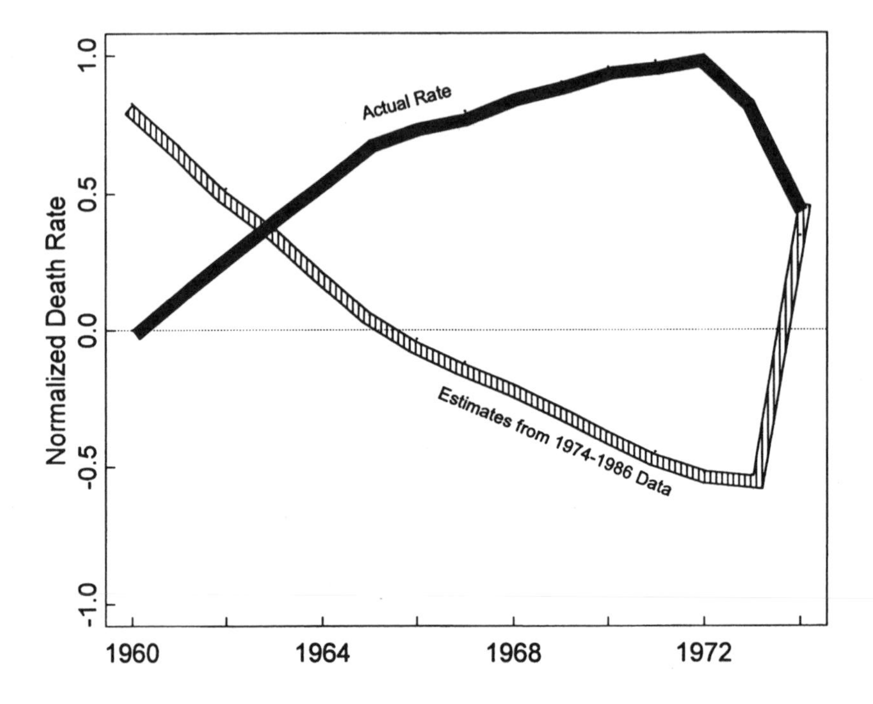

Figure 3.9: 1960–1974 COD Estimates Based on 1974–1986 Data A Reverse Extrapolation

From	To	Coefficient
1960	1986:	0.831
1965	1986:	0.790
1974	1980:	0.502
1974	1982:	-0.117
1975	1986:	-0.590
1980	1986:	-0.829

Table 3.3: Correlation of Speed and Death for Various Time Periods

By choosing suitable intermediate years it is obvious that almost any value desired for the correlation could be found. It is easy to obtain "evidence" that speed increases, decreases, or has virtually no affect on the highway death toll. Just imagine the supporting words in the arguments and in the commercials that pick up on study results:

- From 1960 to 1986, 0.83: "Over the long term it has been conclusively demonstrated that speed is a killer."
- From 1974 to 1982, –0.11 (nearly zero): "Modern technology has made today's vehicles so safe that speed is no longer a factor in the highway death toll."
- And finally, 1980 to 1986, –0.82: "We have proof that the driving of these superb machines at their full capabilities is actually making our highways safer."

By now it is evident that the information (the variables) we have looked at in our attempts to "explain" the number of annual occupant deaths in car accidents are at best incomplete. It seems at this point that we have no idea of what we are trying to do or how to go about it.

What we have here are two sets of data for identical variables but taken during two different time intervals. In looking at the data and at some of the analyses, the changes or differences between the two sets should not be very surprising since we were aware of facts and events that were not in the data but that affected them in significant ways: interstate highway construction, the gasoline crisis, small car production, and seat belts, to name a few. Without this knowledge, how could the apparent conflicts in the data from the two time periods be interpreted? Does higher speed result in more deaths or fewer? In a true research endeavor, such additional information is seldom available. We are never in a position to know that all possible factors have been included in the measurement and data collection process. Conflicting data is a common, even expected, occurrence. In light of this trivial example, it should come as no surprise when each week's health bulletin contradicts the one before it. It's a shame that the health column writers consistently ignore this.

Occasionally, good research results in knowledge that can be used rather than just more detailed correlations. When this happens, the reports need no embellishment to make an impact. One such study hit the media while I was writing lesson 2 partially putting to rest (or at least shedding much light on) a long-standing controversy. Actually, so much time had passed that most of the controversy was long forgotten. In this case, the question was "How much X-radiation is safe?" That X-rays can cause cancer has been known for years, but exactly what constituted a safe dose has never been satisfactorily determined; it is a typical case of inconsistent data from different studies. Standards for acceptable levels of radiation were set by heeding the most pessimistic of the available studies, the common conservative approach. The new report claims that a rather uncommon gene, present in some women, makes them about 1,000 times more susceptible to X-ray-induced breast cancer than other women. Here is a real example of the "brown-eyed people" we worried about in heterogeneous samples. If the new study is correct, it's no wonder that conflicting data have been collected in the past. How many sample populations had significant numbers of this more-sensitive type person? Now that a hidden factor has been uncovered, a large block of "intrinsic cussedness" in the earlier studies should be removed. At last, it is easy to see how those early results of X-rays and breast cancer would have been even more confusing than our car death data and the conclusions about speed. There is no way of knowing how many critical gene subjects were involved in any particular test or sample. This is one of those rare cases in which someone stumbled on a previously unrecognized variable at a time when the topic is still of interest. This is also a good reason to strive for heterogeneous identity in the sample selection, and why one can never be certain that it is obtained.

There is an even more important lesson here. When studies are continually inconsistent, e.g., cancer and estrogen, electric fields and leukemia, and on and on, one thing is clear: Either the data or the analyses are wrong. Most likely, in the researcher's jargon, there are unknown confounders. There is at least one factor that is not being considered. In this example it was the rare gene. One of the pitfalls of statistics in research is that it is so easy to ignore the exceptions such as Great Aunt Sophie in lesson 1 who "always did that and lived to be ninety-eight." We never expect a "100 percent" result, we live with fractions, seldom trying very hard to discover why the "undesirable" fraction was undesirable. This also underscores the need for multiple tests of the same type. As well as the need to show reproducible results, careful reviews of many tests can give clues about unknown confounders.

Notes

1. U.S. Department of Commerce, Bureau of the Census, "Motor-Vehicle Accidents—Number and Deaths: 1960 to 1981," Table 1061, *Statistical Abstract of the United States 1982–83* (Washington, D.C.: Government Printing Office, 1983).

2. U.S. Department of Commerce, Bureau of the Census, "Motor-Vehicle Travel, by Type of Vehicle and Speed: 1960 to 1980," Table 1071, *Statistical Abstract of the United States 1982–83* (Washington, D.C.: Government Printing Office, 1983).

3. Department of Commerce, "Motor-Vehicle Accidents."

4. U.S. Department of Commerce, Bureau of the Census, "Motor-Vehicle Registrations, Factory Sales and Retail Sales: 1960 to 1981," Table 1060, *Statistical Abstract of the United States 1982–83* (Washington, D.C.: Government Printing Office, 1983).

5. U.S. Department of Commerce, Bureau of the Census, "Domestic Motor Fuel Consumption by Use: 1960 to 1980," Table 1072, *Statistical Abstract of the United States 1982-83* (Washington, D.C.: Government Printing Office, 1983).

6. Byron W. Brown Jr. et al., "Nonparametric Tests of Independence for Censored Data, with Applications to Heart Transplant Studies," in *Reliability and Biometry: Statistical Analysis of Lifelength,* F. Proschan and R. J. Serfling, eds. (Philadelphia: Society for Industrial and Applied Mathematics, 1974), pp. 327–54.

7. Peter W. F. Wilson, Robert J. Garrison, and William P. Castelli, "Postmenopausal Estrogen Use, Cigarette Smoking, and Cardiovascular Morbidity in Women over 50," *New England Journal of Medicine* 313, no. 17 (October 1985): 1038–43; and Meir J. Stampfer et al., "Prospective Study of Postmenopausal Estrogen Therapy and Coronary Heart Disease," *New England Journal of Medicine* 313, no. 17 (October 1985): 1044–49.

8. John C. Bailer III, "When Research Results are in Conflict," *New England Journal of Medicine* 313, no. 17 (October 1985): 1080–81.

Lesson 4

Fancy Footwork

At this juncture in the discussion you now have a reasonable understanding of what statistics is designed to convey. The vast majority of the work of researchers involved with statistical studies is conducted using the tools and techniques mentioned in the preceding pages. Certainly there are many variations of what you already know (or have at least been exposed to) but all the basic ideas and concepts surrounding statistical data have been presented. From this point on, the statistician's handbook tends to be devoted to elaborate and detailed methods of coping with special situations or constructing ever more profound tests about hypotheses and confidence bounds. Elaboration of the technical detail has been going on for so long that, by now, the rigor of the mathematics is close to impeccable. One might well wonder then, if the math is so clearcut, why do so many studies go sour? Well, as in many fields, although the tools are good, the users may not be. Statistics may be a science, but using statistics is mostly pure art. In other words, it is extremely subjective.

Recall the story of the 1936 phone book and why it made for a very biased sample. Since then, much effort has gone into that topic and whole volumes are now devoted to figuring out how to pick samples.[1] There are random samples, cluster samples, multistage samples, and on and on. And, as you might expect, as the selection methods get more involved and elaborate, so do the methods of analysis. But don't worry, the basic ideas never change. As long as the researcher remains honest, whether he flips through a 1936 phone book or spends 1,000 hours of computer time, in the end, a sample of something is selected and a method of analysis provided.

In practically all cases then, the causes of real problems with samples must be similar to those mentioned earlier, such as a bad choice for the listing (population) from which to choose. Similar statements apply to many of the other advanced and specialized tools. Legitimate critiques focus primarily on subjective choices made by the researcher rather than the tools that were used. And the subjective choices are numerous.

Just consider sample selection. As you should suspect by now, simply deciding who or what to measure is hardly straightforward. It involves at least the following:

- Determining the variables or characteristics of interest.
- Deciding which members of the population have those or other characteristics.
- Determining if there are other characteristics of interest.
- Figuring out if the characteristics are determined subjectively or objectively. (Is the yea or nay a matter of the subject's memory or something verifiable by outside sources?)
- Calculating how reliable the measurements will be with these variations.
- Estimating how stable the population will be over the time involved (i.e., how many will change their minds over a given number of days, weeks, months, etc.?).

Notice that any or all of these measures may have to be determined by the researcher's best guess (although one *could* do a pilot study), which is, of course, subjective. In this context, it is not unfair to say that almost all statistical studies are subjective in nature. Then, to this pool of subjectivity in selection, we often add arbitrary choices of confidence and confidence bounds or what significance level we will accept as meaningful.

If you are participating in a craps game will you bet for or against the shooter if his number is nine? What if it is ten? You know, or you can look up the chances of throwing either, just as you can calculate or look up the significance level, the chance of being wrong, for your test result. The chances for either are objective, they are simple numbers. The subjectivity is yours in deciding what level to live with or which bet to make. If all of that sounds very unscientific, it really isn't, it's simply the nature of statistics. It is our own misfortune if we lose sight of that fact.

Enough cautions for now. In this lesson, while exploring some refined procedures, we will enhance our feel for the basics with a few ramifications and more illustrations of the ideas that have already been presented. These new procedures will introduce some advanced techniques in a logical fashion and you will see that they are simply more sophisticated (fancier or more technically complicated) ways of accomplishing the same results. This does not mean they are redundant. Certain techniques are "best" applied to certain types of problems for reasons that may or may not be obvious to the layperson.

The car accident data of the last lesson were a little unusual for statistics in that they could be plotted in nice simple, smooth, and continuous lines. It is far more common for the data not to lend themselves well to that sort of presentation. Instead of smooth lines, we wind up with data points strewn all over the place. Such graphs are very properly referred to as *scatter plots.*

An illustration of one of these is available from the data I used to conjure up one of the supporting tidbits for the deadly equator story. Recall that I claimed the death rates within the United States were lowest in Alaska, generally higher in the southern states, and peaked in Florida. I'll show these data, death rate versus state, in a moment. When I do, there will be two kinds of rates shown that need to be explained first.

In data of the kind wherein the value of one variable is compared over many similar but distinct entities—in this case, death rates over many states—there may be unique characteristics of the entities that affect the result. (Well of course there are, or we wouldn't be comparing them.) Perhaps I should say, there are factors that artificially affect the result; that is, something that is really irrelevant, to us, may be affecting the result. For the deadly equator study, the ages of people living in the various states might have one of these artificial and undesired effects. (I mentioned that age was important with regard to Alaska and Florida.) It is often possible to adjust the data to compensate for the unwanted effects or biases (another word for your vocabulary: *bias*). Of course, what is unwanted or what constitutes a bias is often about as objective as my taste for rare beef. For the case at hand, if I were studying deaths due to heart disease on a state-by-state basis, I would probably consider age a bias because I know that is a factor and I presumably know all about it. I might want to adjust for age so that I can look for other factors. The age is just a confusion item that I can get rid of. On the other hand, if I were writing some sort of insurance coverage for people on a state-by-state basis, all I would care about is the overall death rate, not about why it is different from place to place. The only "bias" I recognize is the name of the state. Nonetheless, by way of illustration, we will do an age adjustment for the state-by-state death rate information.

The U.S. government publishes state populations broken down by percentages within age groups, for example, 12 percent are less than five, 9 percent are five to ten, etc.[2] There are also data available concerning the national average death rates within those same (or similar) age groups.[3] Using these data it is possible to construct an expected death rate, based on the national numbers, for the age groupings that appear in each state. We can then claim that, if all the people in the state behaved like the national average, the death rate for the state should be this, our calculated age-adjusted rate.

Once we have done the adjustment, any differences between the real death rates and the adjusted ones must now be attributed to factors other than relative age. In other words, the age-adjusted death rate for each state is what we would expect if all the people in that state behaved like the national averages when it comes to dying. It is just a gimmick that highlights small differences between arbitrary groupings that may be masked by obvious uninteresting differences. In this case, differences in death rates between states might be overshadowed by unique age groupings in them, e.g., a larger population of older people in Florida that skews the state's average death rate across all age levels.

Adjustments of this nature are carried out all the time. Some may wonder what is meant when a study reported in the local paper points out that "Heart disease is still the leading cause of death even after adjusting for age and gender." The statement means simply that the data were adjusted to compensate for, or to remove any biases which might be attributable to, age or gender.

Adjustment calculations are often rather fuzzy. In our case, we had to interpolate (a calculation to guess at values between the entries in a table) to get estimates of rates within age groups that would match the groupings for the state population data. Further, the state population data are only census bureau estimates, with errors of at least 1 to 2, or more, percent. The death rates within groups, even before we interpolated, were again somebody's estimates with errors of their own. What we actually did was use data from two different surveys, population and group death rates, modified some of them, and then compared them to data from yet another survey, average death rate within a state. I wouldn't expect any of the results to be within 10 or 15 percent of the real numbers in the best of cases. It is useful to keep these thoughts in mind when reading or hearing about studies that "adjusted for" various confounding factors.

Now let's look at the scatter plot of death rates and age-adjusted death rates. These data were published for the years 1990 and 1991. For this picture, figure 4.1, I lined up all the states, along with the District of Columbia, in decreasing order of average lifespan or longevity. This is a scale ranging from 77.02 years for Hawaii to 69.20 years for Washington, D.C., a variation or spread of eight years. This is more than twice the total variation reported by the United Nations for all developed countries in the world, except the (then) Soviet bloc, in 1988. That includes Japan, Israel, and all the highly touted Mediterranean countries. If you toss in the Soviet bloc, the range only goes to nine years. *In the absence of further analysis,* it indicates that if you are born in Hawaii you can expect to live eight years longer than if born in Washington, D.C. Why do researchers trek the world looking for reasons for something when the greater variability is here at home? Who would have thought that life expectancy was so dependent on the state in which you live? Although exact numbers don't appear on the chart, we can gain a sense of the range of ages shown for each state. With the states lined up and named along the bottom, the two kinds of death rate for each state are shown with little white squares for actual death rates and little black triangles for the age-adjusted rates.

I like scatter plots, they quickly give you a feel for how the variables change from one extreme to another. Trends up or down are instantly apparent and otherwise hidden patterns of behavior may simply jump out at you. All of this is good, but we must also learn to take second looks for finer points that may not be so obvious. This plot is a good one to practice on.

First note that, as I claimed, the death rate is lowest in Alaska and highest in Florida (primarily because of the age of the general population is lower in the former and higher in the latter). Also, there is no significant tilt in the scatter, that is, if we drew an "average" line across the picture, it appears that it would be flat and horizontal. Longevity and death rate appear to be pretty much inde-

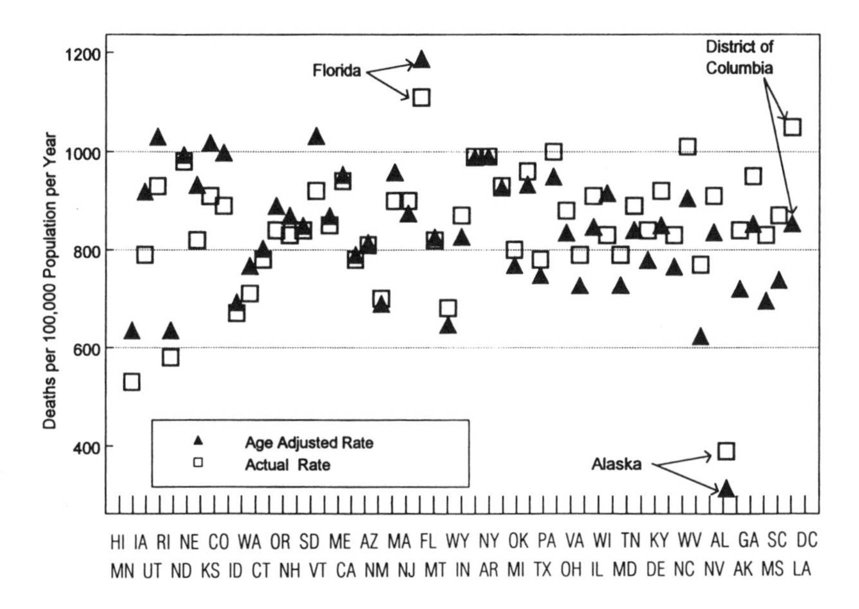

Figure 4.1: Death Rates vs. Average Life Span for U.S. States

---

pendent. Even though that is far from a rigorous test of dependency, it does support my earlier statement that longevity and death rate are not the same thing. As you can see in the top center of the picture, the older people moving to Florida, and so giving that state an average (it's in the middle of the plot horizontally) longevity, also creates that highest death rate. And Alaska, with the lowest death rate, again because of the population's age, has nearly the shortest life span. So, death rate and longevity really have very little to do with each other.

Now, because the age-adjusted rates were calculated on the basis of the national average rates of death within age groups, a distinct pattern can be found in the diagram, but we really have to look to find it. Let's do that.

If we plotted a point for the national average death rate, it would fall about in the middle of the whole picture near Montana (MT), the way averages are supposed to behave. States that lie to the right of that central point have age-adjusted rates that are smaller than (fall below) the actual rates. Notice that on the right side, the little triangles are below their corresponding squares. That's in keeping with the actual lifespans being shorter than those calculated, people there die sooner than expected. In other words, the shorter lifespans appear to be consistent with the higher-than-expected death rates. (Are these independent of one another or not?) Conversely, to the left of center, longer than average lifespans occurs. The adjusted death rates are larger than those actually found. The triangles are above the squares. Now, rather than right smack in the middle, the transition of triangles from below the squares to above them occurs slightly right of center, about where Arkansas (AR) is positioned. See where the squares and triangles fall nearly on top of each other? Have we identified a pattern, a systematic relation between the two kinds of death rates and longevity? More about this in a bit. For now, just take this little exercise in pattern detection as a training session in how to read the pictures sometimes generated by studies. (If you gain nothing else from this book, I hope you develop a love for pictorial data.)

Did you notice Alaska down there in the corner all by itself? Data points like that are called *outliers,* for obvious reasons. Identifying and verifying (or justifying) outliers is an entire subject unto itself. There are even a number of tests that will help decide if one or more points are far enough from the crowd to be called outliers. Naturally, where you set the limits when using those tests is subjective. Outliers must always be examined. They may be simply wrong pieces of data, explainable or unexplainable exceptions, and they may provide clues for lucrative further research. As we claimed earlier, Alaska has that very low death rate simply because the population is very young relative to that of other states. That's confirmed by the calculated age-adjusted rate down there in the corner with the real one. It shows that when the death rate was calculated there must have been a lot of young people included, thus weighting the answer to the low death rate side.

The District of Columbia is pointed out on the chart, not only because it has the shortest lifespan, but it also appears to have the greatest difference between actual and adjusted rates, the former being significantly higher than the latter. I offer no explanation for these observations. They could be quirks in the data of one of the surveys or maybe the Capitol City is really an unhealthy place. It may be that there is grist here for another study.

This little diversion about the District of Columbia could very easily be total nonsense. However, it does serve to make another point: whenever items are lined up on some continuum, it is an obvious but often forgotten fact that two of them have to be at the ends of that continuum. If one end or the other is deemed to have some negative quality, real or otherwise, then the poor item occupying that last position is doomed to criticism and disparagement. It does not matter if in fact the difference along the entire continuum from one extreme to the other is negligible, that detail will be ignored. Simple placement at the bad end is all that counts. A significant event of this type occurred in 1975 with the publication of an *Atlas of Cancer Mortality for U.S. Counties: 1950–1969* by the National Cancer Institute. In this case, several locations in and around the State of New Jersey happened to lie at the high end of the scale for cancer incidence. The term "Cancer Alley" was soon applied to the area and numerous debates and studies were begun. And as it turned out, all was for nothing. The difference between the cancer rate in the so-called Cancer Alley and the national average was of course shown to be minor when finally put into perspective. That is, if seen on a scatter plot, those places would simply have fallen near one edge of the "cloud" of points. They were not even real outliers. No matter how small the total difference may be, if we force an ordering on the items in a study, then one of them is, by definition, the best or the worst of the lot, and that resulting nametag is all that people will remember.

Now, in spite of all these exciting points I have raised, it's likely that the scatter in figure 4.1 still looks to many as though someone simply bumped the salt and pepper. However, with a little practice anyone can learn not to throw up their hands so quickly but to begin thinking of other ways to look at the data. For example, the white rectangles and the black triangles have that peculiarity we noted and explained above, their relative positions reverse as we move from left to right across the picture. Is there anything else going on?

One way to get a quick answer to this question is through another visual display. Plot one of the variables against the other: say, the triangles against the rectangles. To do this in a sensible manner, the rectangles have to be ordered (here we go again with an ordering). The result appears in figure 4.2.

Now, here is a scatter diagram with a real message. Isn't it obvious that the points follow a trend? To help visualize this, I have positioned the linear regression line (remember those?) on the plot. This is a neat visual of the uncertainties associated with a regression line and any predictions that might be made from it. Predictions

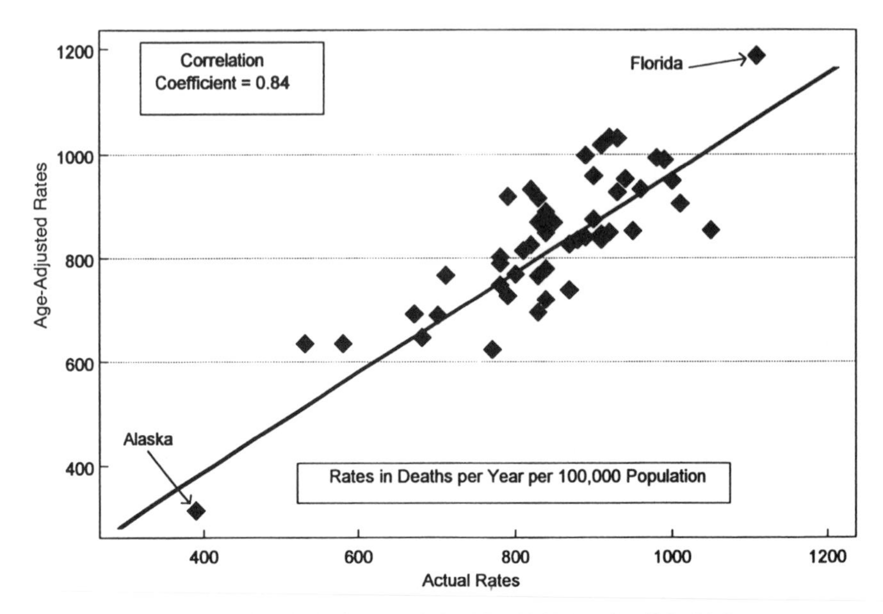

Figure 4.2: Age-Adjusted vs. Actual Death Rates for U.S. States

would of course fall right on the line, but the real data seldom come close to the line. And, in this case they have a shockingly high correlation coefficient, 0.84, as noted on the drawing. This makes the profound statement that the calculated or estimated death rates tend to be similar to the actual death rates.

We see that outlier Alaska again, way down in the left corner this time because of its low death rates. Now Florida appears way in the upper right corner, the antithesis of Alaska. Would Florida also be classified as an outlier? How about the District of Columbia? In general, there is room for much discussion on what constitutes an outlier. In spite of, or maybe because of, intrinsic argumentativeness, outliers can be more fun than three sigma extremes. Depending on your whim, you may use them to either support or undermine your colleague's position.

Given the way the data in figure 4.2 were derived, and the fact that the plot is simply two different measures of the same thing, death rate and age-adjusted death rate, there seems to be little of real interest in the picture. It is simply a more dramatic demonstration of the differences between the two death rates, actual and calculated, than was achieved in the first picture. In addition, it is a good representation of a scatter plot with a definitive trend. It does not, however, say anything about cause and effect. It doesn't help us to explain anything about why people die. If our friendly Neanderthal had plotted the lengths of nights against daytime temperature, his result would have been similar. Nevertheless, researchers trying to show a causal relation between two things would be ecstatic over such well-behaved data. The high correlation, of course, would convey the same ideas without need for the actual plot. But, the special cases of Alaska and Florida, had we not been aware of them beforehand, would have probably gone undetected without the diagram. And my significant remark about Alaska in the deadly equator scenario would have been omitted. See, it really is important to look at the data. Scatter plots can be useful; it's a shame that researchers seem to ignore them. You rarely see them in a report. This could indicate a lack of interest, poor training, or the fact that *good* software for generating them has not been widely available until recently. Who would even consider making a scatter plot by hand? Note, by the way, that the ordering of death rate forced those two states, Alaska and Florida, to occupy extremes on the continuum and in this case, the extremes are so far removed from the rest of the points that special investigation would be warranted if we did not already know the reasons. The remarks above about end points seldom being significant do not apply here. The ends of this continuum are strikingly separated from the cluster of main events. As for the District of Columbia, all this picture does is verify that, yes, it has the largest difference between the calculated and actual death rates. It is the farthest removed from the regression line. Is this an indication that the inhabitants of the area are peculiar in some way?

While we are on the subjects of scatter plots and regressions, it may be useful to illustrate a fundamental difference in the thought processes of different kinds of analytical professionals: scientists and statisticians. Either of these two types might come up with a plot similar to that in figure 4.2, and both would draw and present it in essentially the same way. The scientist, especially a physical scientist, might be rather discouraged with the result, or at least believe that there is still a lot to be learned. He might conclude that a considerable amount of dispersion occurs around that line. Thus he must still be doing something wrong. Remember, the physicist strives for zero error because he deals with exact relations. If the error cannot be driven to zero, or to within known experimental error, he withholds all conclusions. Conversely, the statistician, as we have indicated, would be enthralled. His regression line with its correlation of 0.84 tells him that his reduction in uncertainty is 71 percent (0.84 squared times 100, the coefficient of determination). That's very satisfying in the statistical world. Figure 4.3 is a representation of these two modes of thought associated with the same result.

In lesson 2 I claimed that the researchers liked to use straight lines in their regressions. Although there are any number of "wavy" line alternatives, they are "nonlinear," which translates into difficult, even in this computer age. The problem is easy to describe. A straight line can be drawn at any angle and made to be any length, but it is still a straight line. Once we decide that our data (think of the scatter plot of figure 4.3) can be suitably represented by a straight line, most of the real work is done. Analysts have used straight lines for a long time and developed many tools to go with them. One can quickly get confidence bounds, risk factors, or other numbers relevant to the problem at hand. Straight lines have been beaten into the ground, so to speak. So, once we have one, we know everything we ever want to know about the data. This wealth of information available, once the straight line has been picked, has been accrued through decades of hard work. If we add a curve to our line, all bets are off. Most of that wealth of information is unusable, we have to start over. To make matters worse, while a straight line is a straight line is a straight line, the same is not true for curved lines. There is no end to the number of possible curved lines. So, at first glance, if we decide to go nonlinear, we are breaking new ground.

In the realm of medical studies there are a few exceptions to the nonlinear enigma. One is especially popular, and there would appear to be two reasons for its attractiveness. The first is a plausibility argument with considerable intuitive appeal, the second is a mathematical trick or transformation that makes the use of this particular wavy line almost as simple as that of straight lines.

This curve is one that has received a lot of attention from economists, biologists, naturalists, and others generally interested in growth patterns. Stop to think for a moment about the way things grow, especially a new collection of items: new ant colonies, the grass on your lawn, a new franchise business, etc. The growth of such things is characterized by a slow start followed by a spurt lasting for perhaps some time, and finally a flattening out or leveling off when the things mature or the market is fully penetrated. Competition may also contribute to the leveling off. At this point, the reasons are of no interest, the common shape of the growth pattern

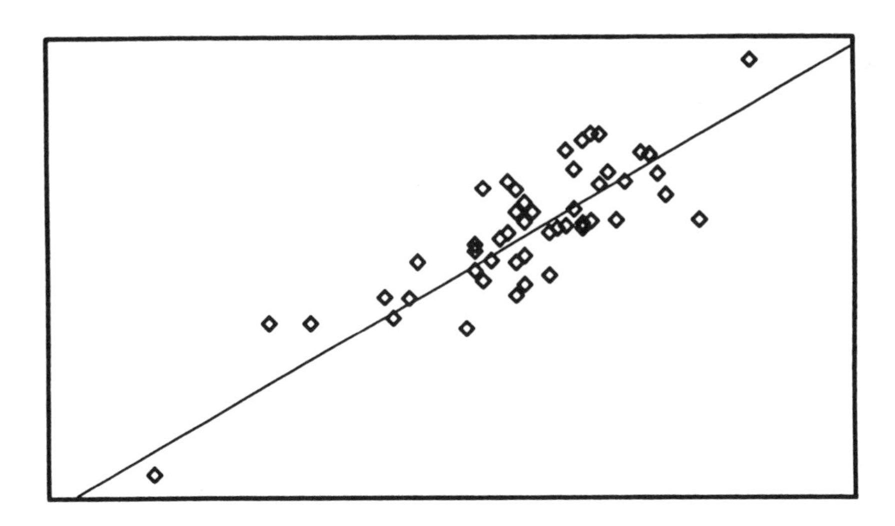

The Scientist's Pen

The Statistician's Brush

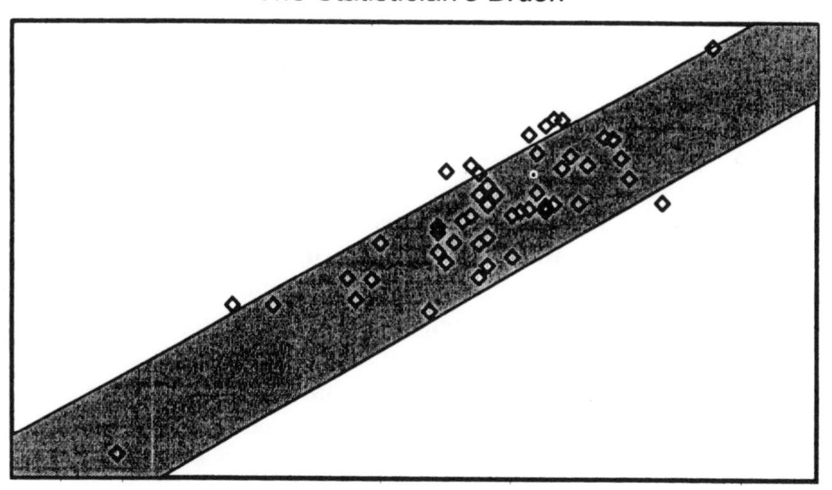

Figure 4.3: Perspectives

---

is what is important: the slow start, rapid rise, leveling off; it can be visualized as something like an S-shaped curve. The S is tipped forward so that as we go from the bottom to the top, there is no need to travel backwards or to the left for a time as the S actually does. On these S-plots, the label on the left side of the picture is the size of whatever it is we are tracking, like the height of our grass by next

Saturday. Along the bottom edge of the picture we find labels indicating the quantities of things used to permit, or even encourage, the growth. Plain old time is a common one. For our lawn it may be a mixture of time, water, fertilizer, and sunlight. For example, if we do nothing but let time pass, the grass may not even need mowing next week. But, keeping up with the neighbor, we add a little fertilizer and so we can count on mowing next week. Then maybe we water a few times, there goes all of next Saturday. That old S-curve is beginning to climb. If we're really into it, we may have to mow twice a week, but probably never more than that no matter how much we water and fertilize. The S-growth pattern has run its course and flattens out. Now if we insist on adding ever more fertilizer and leave the sprinkler on all day, every day, that brand new lawn will disappear. The top of that S-curve will turn straight down and, in a short time we're back to dirt (or mud). Even though the S-curve is a lot fancier than the straight line, it is still rather simple-minded; it has no drop at the end to follow that dive into the mud. And so, it does not permit studying real cases where such a drive might happen. Given today's computers, it should be straightforward to find a curve that would include that drop, but that would be a new nonlinear curve, one not beaten into the ground. Stick with the S-shape and its simplifying transform (which we haven't talked about as yet) and consider the study to be over when it flattens out.

A while back, someone associated with the medical world recognized that the S-shaped growth curve might also be useful as a description of our chances of catching something undesirable. Since disease is usually the uncommon circumstance, the chance of getting one is typically small. On the other hand, as we become exposed to, say, people who are already infected, our chances of getting sick rise very quickly and may approach a certainty. We all know that when little kids start school, we can expect a rash of colds. Rather than water and fertilizer, the driving force along the bottom of the picture is now some collection of risk factors and exposures to them. Just as the fertilizer and water help push up the grass, more exposures to more risk factors help increase our chances of contracting disease. At least that's the rationale. Note that already, at this early stage, we have fallen into using descriptions that sound as though we have accepted the cause-effect relation even though, deep down inside, we know that that relation cannot be statistically proved. See how easy it is to fall off the edge of reason? All we are doing is trying to set up a model that describes and quantifies something in a way which we hope to find useful. It's so easy to let the model, this S-shaped curve, carry us away. A well known statistician has cautioned against this by insisting that "Models should always be used, but never believed." That is his way of saying that you imposed the model on the data to help you understand and, though it seems to fit, if questions arise, believe the data, not *your* model.

This magical S-curve does have an official name: the logistic curve. The logistic is actually one of a small number of wavy lines used in medical research. All have that general S shape, or a part of it, but the non-logistic curves are more difficult to use. Even the gurus readily admit that simplicity of form is the major criterion. That is not all bad since we agree, up front, that we will never believe the model. If that is so, why spend time and energy constructing it? The model is not unlike the Hollywood set for a Western. It's sufficient to build the facades; the concept is captured even though reality is totally lacking. However, when the researchers' movies are shown, people tend to behave as though seein' is believin'. Recall that before they realized what was happening, the company managers at Hawthorne believed that more lighting *always* increased production. We will explore modeling in some depth a little later.

The multisomething regressions we performed earlier allowed us to make guesses about numbers—how many car occupants would die, for example. By playing the game in appropriate ways, our lines could predict numbers for anything about which we had data. We could, for instance, predict the number of accidents, the gallons of gas consumed, and so on. Not so for the S-shaped curve. This logistic model is more limited in that it permits only a guess at the chance that something will happen rather than the number of times. The label on the plot is and always will be chance, and the curve is constrained to values between zero and one, those of chance or probability. Take a look at the sample curve in figure 4.4.

The thick arrow along the bottom with the rather nondescript label "Increasing Exposure to Risk Factors" is truly the best I can do for a simple explanation of what is happening along that horizontal axis. I could put numbers along there: as this particular picture is scaled, they would run from about –7 on the left side, to 0 in the middle, and +7 on the right. The –7 would correspond to a zero chance of being afflicted because it is at the bottom of the S-curve. Now, tell me what it means to have that obviously desirable exposure of –7 and its corresponding zero chance of affliction, as opposed to having one of 0 or +7. For instance, if the curve has to do with the chances of

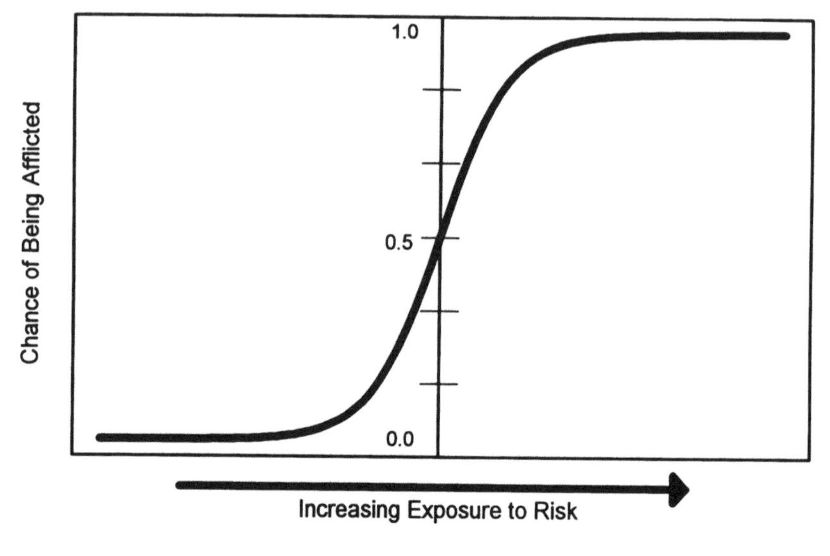

Figure 4.4: The Logistic Curve

---

tooth decay, tell me how much candy I have to eat before my teeth fall out. Can I brush my teeth? How often? With what? Is a dentist visit permitted? As I said, for now, the arrow and simple label are adequate. The only information there, and all we need to have at this point, is the idea that greater risks imply greater chances of becoming afflicted with whatever it is we are trying to describe.

All those questions about brushing and so forth are not as facetious as they appear. In any multirisk-factor study, those are exactly the kinds of things that get plowed into the number scale that will eventually appear along the bottom of the picture. Think about it: Aren't all of these factors supposed to be important in the general problem of tooth loss? By way of illustration, let's make up a study about why teeth fall out. We will plan it as though it were actually to be carried out with real people.

First of all, in picking people as subjects for this study, we must select from only those who have already lost all their baby teeth. Obviously, those loose precursors of permanent teeth could mess up the results. To keep it simple (and totally naive) we will make believe that all the test subjects are absolutely honest, reliable, and trustworthy. What we want to do is have their teeth examined at the start of the study. We must know exactly how many teeth are in each mouth. All participants must agree to the length of the study. Let's make it five years. Obviously, if we determine, by methods not discussed here, that we need 1,374 people for reasonable confidence in the results, we had better pick a few extras to allow for uncontrolled departures (dropouts) during the five-year period. The object of the study is to estimate the importance, in the general problem of tooth loss, of each of those items we questioned. At the end of the five years we count everyone's teeth again and register, for each subject, a yes or a no for lost teeth or no lost teeth. I'll explain a little more below.

When the subjects for the study are first interviewed, the questions to be addressed, and the kinds of answers to record, are as follows:

1. Do you brush?	0 = no
	1 = yes
2. Times per day?	0, 1, 2, 3, 4
3. With what?	0 = don't
	1 = plain paste
	2 = fluoride
	3 = lemon juice
4. Visit dentist?	0 = no
	1 = yes

This may seem simplistic but it represents a very complex matter. What you ask your subjects, and the way they reply, must be definitive. There can be no ambiguity about any question or its answer. Maybe, or both, or I don't know, is not permitted. If people insist that such is the case, they don't belong in the study. Putting together a questionnaire for your study is not a trivial task.

Now, again assuming ideal subjects, each of them can be assigned a four-digit code corresponding to their four answers. For example, someone who brushes (1), twice a day (2), with plain paste (1), but doesn't visit the dentist (0), would get the code 1210. Even though we shouldn't anticipate, the poor guy with an all zero code, according to common belief and commercials, should have another zero at the end of the five years representing the number of teeth left in his mouth. (My mother-in-law is a zero-code type, but still has thirty-two pearly gems at age eighty-five. Pardon the interruption, but I just love these exceptions.) Anyway, after the final exam, as we just suggested, another digit is added to the code, a one for no tooth loss or a zero for some. All the coding is for the computer, but it also helps to keep us focused.

If we had examined a lot of people, it is likely, or at least we hope, that for any given code, at least one person with that code will have lost at least one tooth. Otherwise we would wind up estimating a chance of zero for tooth loss and such an estimate is fraught with problems from the obvious question "Why that code?" to "Do it over with a larger sample." Well, for each code, we count up all the people who lost teeth and divide by the number with that code and—voilà! We have a guess at the chances of losing teeth for each and every code. Well, almost every code (you could not have a code starting with 1 then 0, meaning yes, I brush, zero times a day). Now it's time to bring on the computer again, dump all of the numbers in and, if we have the right software, out falls a neat little equation for an S-curve. In essence, what the computer did was rank the chances of losing teeth for each and every code, "plot" them and draw a "best fitting" S-curve through the points. The equation for the curve lets each of us (with our own personal code) find our spot along the bottom of a figure like that in 4.4, look up at the curve and read the number at the left to tell us our chance of losing teeth in the next five years. That's the way it's really done, and if we don't buy this scenario, maybe we should think twice before we buy the next health fad product.

Is there a catch here? The answer is yes, in one sense. Remember the term *correlation coefficient* that we used a while back? Recall, I said that it neatly hides all the cases that don't fit the thesis, like the warm winter days that plagued our poor Neanderthal, or the fact that practically none of the points in a scatter diagram fall on the regression line? Well, the equivalent here is the use of chance, the vertical scale in figure 4.4.

Consider the particular subject code in our tooth study, whatever it may be, that winds up registering as 0.8 on the S-curve and, suppose that 295 subjects had that code. Working backwards, that means that 236 of them (80 percent) lost teeth, fifty-nine, however, did not lose any. The same kind of statement is true for every point on the scale. But isn't this exactly what "chance" really means? What's the message? Well, the point is that, at least in my view, this study has done absolutely nothing. I have known the principles of oral hygiene from preschool days, attaching numbers to toothpaste or lemon juice is not going to change my habits, much less guarantee my placement in that lucky group of fifty-nine. In my opinion, the study fund would be better spent trying to figure out why my code-zero mother-in-law still has all her teeth. Now that's a study to which I would contribute.

Seriously though, this dramatizes the lack of evidence about cause and effect in all statistical studies and the fact that the only thing they "prove" is a need for real research. If any of the factors in codes that lead to the 0.8 estimate of chance are really causes, why did fifty-nine people not lose any teeth? There is nothing in this study to tell us, no more than there is anything here to tell us why 236 people did.

Finally, on to the promised transform, the mathematical trick that makes the S-curve straighten out. If we were reasonably happy with the S-curve that came from the tooth-loss example, and the fact that we could punch in our personal code and guess when our teeth might fall out, you may ask "Why bother?" Well, people who are interested in study results like answers to questions such as, If I brush four times a day instead of three, will I be better off? By how much?

Of course, changing the appropriate number in your personal code from 3 to 4 and plugging that into the formula finds the new spot in figure 4.4. That's fine, but if you ask the question in reverse: "How much more often must I brush to improve my chances by 3?" getting a solution is difficult. The only way is to just keep plugging in different codes until you find the right one—the trial and error method. That's a serious shortcoming with this type of nonlinear regression: You can't solve the equation in the reverse direction as you can a linear one.

Furthermore, because these curves are multivariate, we can find a whole bunch of codes that give the same 0.8 answer. For instance, brushing twice a day with plain toothpaste might be just as good as brushing once a day with fluoride. Finding all the possible codes by trial and error for every different question would not be easy. Life would be simpler if we could straighten that curve with a transform.

There is another motivation for wanting to transform that S to a straight line, albeit a sort of fortuitous one. For mathematical reasons that we shall not address at this point, researchers for years have talked about the odds of something happening rather than the chances. There is a difference. Numerically, chance can only range from zero to one, but as we know, odds can be almost anything from less than one-to-one to more than a thousand-to-one. Chance and odds, intuitively, are practically identical, and of course they certainly convey identical information, but mathematically they are different.

To estimate chance just take a coin, toss it a lot, and count the numbers of heads and tails. To make it interesting let's suppose it's a biased coin. This means that due to the distribution of weight in the coin the numbers of heads and tails will not be the same, or even nearly so, no matter how many times it is tossed. Suppose that in one million tosses, we count 750,000 tails and 250,000 heads. To guess at the chance of tails over heads we ask the question, what fraction of the tosses were tails? The answer of course is 750,000, the number of tails, divided by one million, the total tosses, or simply $\frac{3}{4}$. Notice that the answer is between zero and one just as we promised. To guess at the odds of tails versus heads, the question is, how much bigger is the number of tails than the number of heads? Now the answer is 750,000, same as before, divided this time by the number of tails, 250,000, with a result of 3. So, the chance is $\frac{3}{4}$ but the odds are three to one. Many people are probably well aware of this, but it was necessary to make the point of a mathematical difference in the two everyday ways of talking about, say, lotteries for example. For the mathematically inclined, if P is the chance of tails, and Q is the chance of heads, their sum must be 1. So, $P + Q = 1$, then $P = 1 - Q$, and the odds are equal to P/Q or, $P/(1 - P)$ which is the same thing. One is available from the other, so they do in fact, carry the same information.

Now that we really understand the difference between chances and odds, we can proceed. If we manipulate the formula for the S-curve, which yields chance (P), to calculate the odds ($P/[1 - P]$), we will get a nice straight line just as in linear regression. Because of the type of equation for the S-curve, we must take the logarithm of the odds, but that's a minor detail. Even if we don't take the logarithm, it's much easier to manage changes in the odds than changes in the chance. It's a simpler equation. It is also easy to do things such as compare the odds of Sally Jones's teeth falling out to those of Sammy Smith, in case someone has an interest in Sal and Sam. A note for any vocabulary buffs: the logarithm of the odds is sometimes called *logit*. It sounds as though someone envied the creator of "bit."

Even though chance and odds are basically equivalent, the use of odds has an advantage in certain circles. Looking again at figure 4.4, notice that the S-curve looks flat over almost two-thirds of the picture, part at the bottom and part at the top. Moving along the x (horizontal) axis, or the risk factor arrow, in these areas doesn't much change your chances of things. However, as we will see, the odds in these areas change dramatically. Suppose for a moment that our personal code for dental care fell near the tip of the arrowhead in figure 4.4. The chance of our teeth falling out might be 0.98 as read off the S-curve. Then, if for some reason our code changed just a little bit and we moved slightly more to the right, that same

S-curve might give us a rating of 0.998. Big deal some might say, all the way from 0.98 to 0.998? Well now, don't be too hasty, the odds for those two cases are about 50 to 1 and 500 to 1. (Use the formulas above if you want to check.) Now maybe more people will sit up and pay attention. So, this odds-thing is like a magnifying glass or a very fine scale compared to that of chance. These kinds of numerical differences are undoubtedly the reason gamblers prefer to converse in odds instead of chances. Wouldn't a potential player feel differently about a choice between two horses with odds of 50 to 1 and 500 to 1 than he or she would between two horses with chances of 0.98 and 0.998? Track buffs might want to return to this page the next time they have to hitchhike home.

This numerical difference between chance and odds ends up, rather inadvertently or by default, becoming the substance of dramatic headlines announcing results that may be merely so-so. This can happen in so-called case-control studies in which a group of sick people is compared to a group of healthy people to look for causes of the illness.

When reporting studies, comparisons that are frequently made are similar to that question about attempting to keep my teeth a bit longer by brushing once more per day. Most of us have seen the headlines or heard the newscasts reporting that "The odds of getting Phister's syndrome are thirty times greater for people with blue hair than for those with other colors." This is not just eye-catching, it will make us look in the mirror. It sounds as though our chances of being attacked by Phister's syndrome are thirty times greater than those for "normal" people if we are unfortunate enough to have blue hair. If that doesn't send us to the doctor or hair dresser, nothing will. Even if the report is reliable in the sense that it is based on a fair number of good statistical studies the information we just received is zero! We have been told nothing about our real danger of an encounter with Phister's malady. The reason for this dispiriting fact is based on another mathematical quirk and goes something like this:

In our study about teeth falling out, we said we would estimate the chances of losing teeth by dividing the number of subjects with missing teeth by the total number of people with the particular code. That's fine. A minor rub is that in order for the estimate to be even close to reality, the total number of people must be quite large, especially if the chance is small. I'm sure few will be surprised to learn that, especially in medical studies, it is not easy to get a large number of subjects with Phister's syndrome and blue hair, for example, or for many other conditions that are studied. Furthermore, in order for the arithmetic to work out properly, our two samples of people, some with and some without the Phister problem, should be equally representative of both those groups, say 1 percent of each. At least we must know what percentage of each. But how in the world can we determine that? We don't know much about these groups—in particular, we don't know how many people out there really have the condition, so how do we know what percentage of each group we have?

Here comes the mathematician to the rescue. Because of some detail in the algebra, it turns out that not nearly as many warm-bodied subjects are needed, and we needn't worry about what percentage if we estimate the odds for two different outcomes, then talk only about their ratio. Therefore, researchers are inclined to find and report odds ratios instead of risks or risk ratios.

That may sound okay, but the rub is that difference we talked about at the racetrack. Although odds may mean something to the gambler and appear to be a very fine scale, that very same scale winds up making gross exaggerations. Back there we compared odds of 50 to 1 and 500 to 1, a ratio of 10 to 1, with chances of .98 and .998, a ratio of 1.02 to 1. Take another look at the chances. The first one says that in 980 times out of a 1,000, the horse will win, the second says that in 998 times out of a 1,000, the horse will win. In the best case the horse is only 2 percent better than in the worst case. It is the same as that for two superhuman ballplayers with batting averages of 980 and 998; the top one is only 2 percent better than the other. The odds ratio however, the one in the banner headline, is not the piddling 2 percent bigger, but 900 percent! ([500 – 50]/50 × 100) Even though the researchers do indeed report results in this way because they have no alternative, and supposedly they understand all this, what about us ignorant folk just reading the headlines? If those numbers for the horses had come from a cancer study, how would you react to: "Drug Slashes Odds of Cancer 900%" as compared to "Drug Reduces Risk of Cancer by 2%"?

That difference between the chance ratio and odds ratio—2 percent greater versus 900 percent greater—is not fixed or constant. It varies depending upon what the smaller chance is when we start. In all fairness, we must point out that if the basic risk, as the chance of disease is called, is small, then the difference between the risk ratio and the odds ratio becomes small. So, for very rare diseases, the two are nearly the same. If we are addressing heart disease or cancer, however, the differences are huge. For example, the overall risk of death from heart disease in the United States in 1993 was estimated at 0.326 or 32.6 percent. We are continually hearing that something bad for us, cholesterol, alcohol, or whatever the latest test substance, will increase our chances of heart disease by some scary number. It is seldom clear whether the news report is talking about relative risk or relative odds. If you think that is unimportant, look at my next picture, figure 4.5. It shows the resulting odds ratio for a wide range of risk ratios when the basic risk is that of heart disease, 32.6 percent. For example, if the increased danger is reported as 9, the odds ratio, it might actually be 2.5, the risk ratio. As you see, the differences can be gross. Now the odds are good that the columnist is not aware of the difference between chance and odds so, when the report was read, he or she probably did not even notice and, when writing about it, simply uses the word most liked. We just don't know.

This artificial inflation of numbers by the odds ratio is only part of the story, and the risk ratio itself is not much better. The basic risk, which is seldom published because it is very difficult to estimate, is an absolute must to know if we want to understand what's going on. Suppose, for instance, I told a person that he could reduce his risk of a kind of accidental death by a factor of 100 by doing something very easy. That sounds like a good deal, doesn't it? Well, by wearing an army helmet whenever he goes outdoors my friend would lower his risk of being killed by a falling meteor by about that much. Now what should we think of this risk ratio of 100? Probably we couldn't care less. The reason we feel this way is that the basic risk is so small, certainly less than 1 in 100 billion, that changing it, even by 100, is of no consequence.

So, I remind you once again, that either ratio, risk or odds, leaves something to be desired. Using either one removes the real meaning. Without knowing the basic risk, you receive no information. You haven't been told anything that provides a basis for a rational decision. Stating these ratios is like saying that "up" is twice as far as half-way up. Even though the risk jumps by 2.5 or the odds jump by 9, you don't know whether you're talking about being hit with a meteor or catching cold. Still, I bet that if you have a blue topping you will call the hair dresser to ward off Phister's syndrome.

A really unfortunate part of the necessity for this odds ratio calculation is that the mathematics that allow the odds to be estimated from small samples also make it impossible to estimate the basic risk (large samples are needed). Technically then, from this study, even

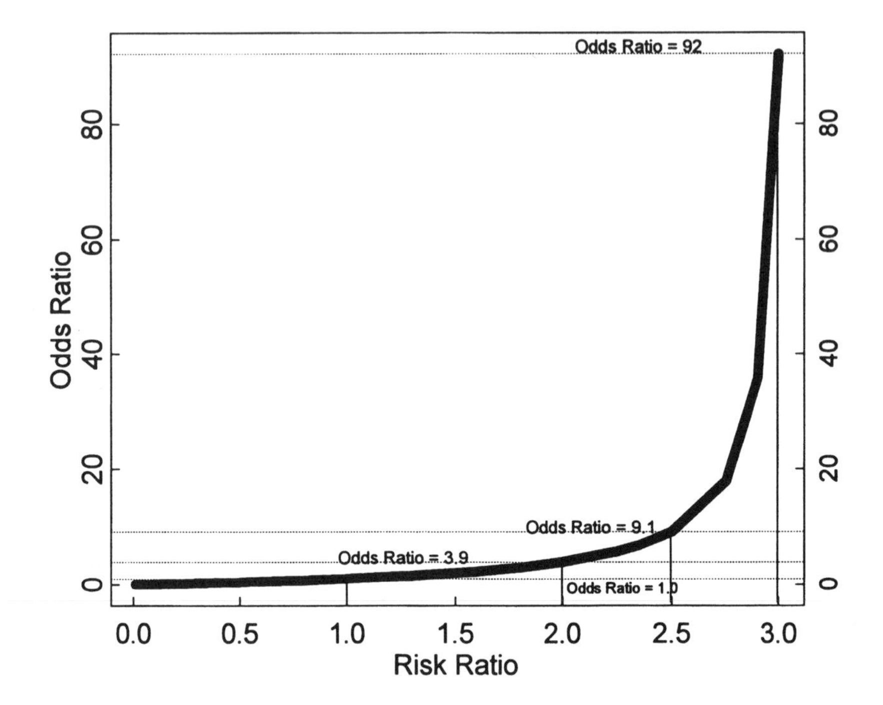

Figure 4.5: The Inflationary Odds Ratio for a Basic Risk of 32.6 Percent (Risk of Death by Heart Disease, 1993)

the researchers cannot know the full meaning of their result. We trust of course, that they have some knowledge of the basic risk from other sources. The rest of us, however, are left on our own to decide if what we just heard is important or of absolutely no consequence.

As a final example of this risk versus odds, consider the chances of dying from any one of the many circulatory problems plaguing society: stroke or heart disease. If you are an American male between forty and sixty years of age, the chances of eventually dying from these classes of diseases are about 0.48 and 0.06, respectively. Now suppose some researcher claims that if you have a mottled thyroid your chances of dying from some circulatory problem will double. On the other hand, if you have a striped thyroid, your chances of dying from a stroke are doubled. The risk ratios for the two cases are identical and equal, 2 to 1. However, the odds ratio becomes 29 to 1 for the circulatory problems and 2.2 to 1 for the stroke problem. Even though the increased risk is the same for both situations, the odds ratios make it appear that the circulatory problem has increased thirteen times more than the stroke problem.

Flash! We conclude this lesson with a bulletin: A few years ago the health news was a real belly-bender.[4] It seems that some physicist in California took it upon himself to look into the power line-cancer controversy (that living in the vicinity of power lines increased the likelihood of contracting cancer) by examining historical data beginning about 1940. He discovered that the use of electricity has increased 2500 percent but the incidence of the types of cancer supposedly associated with electric fields has not changed. Obviously, electric fields do not equal cancer.

Well, the epidemiologists, the people who study these problems, were all over him:

- His records were inadequate.
- There were so many new drugs and treatments for cancer that the results were meaningless.
- There were so many new causes of cancer that the study didn't say anything.
- He could not even show a relation between smoking and cancer with these data.

And so on, but the really jolting objection, as reported in the article, is this: Some experts claim that we don't even know why the types of cancer which have increased (lung, colon, breast, and prostate) have done so, and there is a rationale that indicates that electric fields might cause two of them. These last quoted experts would be happy to claim that the electric fields cause breast cancer, and attribute that increase to more use of electricity. And at the same time they will argue that the increased use of electricity has absolutely no bearing on leukemia. So what do we think now of our objective, rational, nonemotional researcher? Are statistics subjective or not?

That little interruption serves to further emphasize the fact that statistical studies not only cannot prove anything, but that the researchers are painfully aware that they can't as well as of the fact that their study conclusions are fundamentally subjective. Recall our discussions on errors, confidence bounds, and significance: The values we choose to accept are whatever our hearts desire. All statisticians' brushes are wider than the scientists' pens, but whether

I use a small artist's brush or a painter's spraygun, the conclusions I draw are my choice.

Notes

1. William G. Cochran, *Sampling Techniques,* 2d ed. (New York: John Wiley & Sons, 1963); and Klaus Hinkelmann and Oscar Kempthorne, *Design and Analysis of Experiments* (New York: John Wiley & Sons, 1994).

2. U.S. Department of Commerce, Bureau of the Census, "Resident Population, by Age and State: 1991," Table 27, *Statistical Abstract of the United States 1992* (Washington, D.C.: Government Printing Office, 1992).

3. U.S. Department of Commerce, Bureau of the Census, "Deaths and Death Rates, by State: 1970 to 1990," Table 108, *Statistical Abstract of the United States 1992* (Washington, D.C.: Government Printing Office, 1992); and U.S. National Center for Health Statistics, "Section 1—General Mortality, Tables 1–9," *Vital Statistics of the United States, 1990* (Hyattsville, Md.: National Center for Health Statistics, 1990).

4. Report in *Los Angeles Times,* April 14, 1992.

Lesson 5

Esoterica

The questions and problems raised by the subjectivity in routine statistical studies are insignificant when compared to those arising over our next topic. This one is slightly different than those examined so far in that it is not simply another statistician's tool, but an attempt to use a great many of them all at once to satisfy a demand. This lesson represents a slight diversion from our theme thus far, but it will broaden our background in the kinds of games that can be played with statistics. This should add to our objectivity when perusing reports that claim to make significant changes in our lives.

While the kinds of studies we have discussed often deal with serious subjects, e.g., chances of heart attack, we are so used to the on-again-off-again stories (the drug is good today but bad tomorrow) that they have effectively become a frivolous part of our lives. It seems as though everyone jokes about them. Besides the consistent inconsistency of these reports, there is an apparent lack of authority and responsibility, and therefore integrity, of the authors. Sometimes they are specifically identified, for example: Dr. H. H. Hertzhurts of the Institute for More Studies, but often they are simply referenced as the "experts." I don't know who the experts are and I never heard of Dr. Hertzhurts; I'm not impressed. However, when the federal government gets into the act, in the form of the National Institutes of Health (NIH), the Food and Drug Administration (FDA), and so on, I'm more inclined and so are millions of others to pay attention. After all, these experts are paid, with my money, to guard my health and make me live as long as possible, or longer. Now that's an awesome responsibility, requiring, among other things, some awesome tools. Currently, the most popular of these tools is not just awesome, but positively wondrous. It is not just a statistical tool, but a blending of a whole host of them along with a touch of real science. It has been around in various formats for several years now, but is still evolving. It is called quantitative risk assessment (QRA). I classify it as esoteric because there is no way it could be made completely understandable and acceptable to the general public. This is true in spite of the fact that it was created (and continues to evolve) in the public's interest and in response to the public's demands. If we aspire to be practicing statisticians, this is a sample of the kind of real-world problems we might face. The story of QRA, what it is and what it does, or more to the point, what it doesn't do, goes like this:

Some may have guessed by now that QRA is a method for determining how dangerous things like pesticides and nuclear reactors are for purpose of their regulation. This is certainly an interesting problem and worth some real effort but, right off the bat, there are big problems. Here is a trivial example to give some feel for them so we can appreciate QRA.

Forgetting for the moment about exotic things like nuclear reactors, how would we answer the danger question for simple, well-known items? Well, we might say, anything that kills people is certainly dangerous, and obviously, things that kill more people are more dangerous and so these items should have tighter controls. Having blurted that out and realizing what we just said, we might rephrase the last part to read something like: Things that are potentially more dangerous should have tighter controls.

If those last few lines throw you a little, consider what it was that we really said in the first case about things that kill people:

1. That people have died means danger exists.
2. With more deaths there is more danger.
3. As the danger increases, the demand for tighter controls increases.

That sounds good until we apply it to nuclear reactors and airplanes for instance. Airplanes kill hundreds of people in this country every year but nuclear reactors have never killed anyone in this country, at least by any means directly associated with the nuclear aspects. In spite of this, we don't find demonstrators trying to shut down Boeing Aircraft or American Airlines. The point is, potential danger, real or perceived, is often much more important than demonstrable danger, i.e., body counts. Having made this point, we will drop it for now because we do not have to concern ourselves with the dangers of well-known items such as cars and airplanes. Their very real annual tolls are tracked for a number of reasons and it seems that we are all pretty well content with the current bag of safety devices, speed laws, and body count.

Let's continue with the problem of determining how dangerous well-known items are so that we can test our tighter control theory, number 3 in the list above. For this purpose I have looked back over data from some recent years and come up with rough averages for annual death rates associated with eight common things shown below. I have ranked the items from least lethal to most lethal according to the number of people killed each year by them. This allows me to say that for every person killed through the use (or misuse) of the least lethal item, three are killed by some other item that causes three times as many deaths. Of the items I have chosen, the least deadly one may seem surprising at first; it is explosive material. That's right, dynamite, TNT, and so forth kill very few people so, by our own measure, they are the least dangerous. The remaining items line up as follows. For each person killed by explosives:

- three die from electricity,
- four each die from boat, aircraft, and gun accidents (not murder),
- six each die from medicines and poisons, and
- 120 die in motor vehicle accidents.

Now, according to our tighter control theory, dynamite should be sold in showrooms and cars restricted to experts. Don't bother, I know all your arguments. The real lesson here is that, although almost no one is willing to recognize the fact, and even fewer will admit to it, we are all only too happy to accept a large number of dead bodies when the economic, social, and recreational penalties involved in banning the killer are significant. Our love affair with the automobile is the outstanding refutation of those who argue that economics should never be a part of risk assessment. Or, as those advocates put it, human life cannot be equated to dollars. (Some even proclaim this on their bumper stickers!)

If we reflect for a moment on that dollars-versus-life controversy, we soon realize that more than just dollars are involved in these arguments, especially when the object in question is the automobile. If we consider banning cars, we might estimate the total number of businesses closed, jobs lost, increased costs to ship everything everywhere by other means, and so on, and come up with a guess of total dollar cost. However, even if a benevolent group from Mars offered us the money, we still would not give up our cars. But of course, someone will argue, there are other aspects, the social and recreational penalties mentioned in the last paragraph, for which we can't clearly determine dollar amounts. True, ethereal though they may be, these "human values" frequently carry even more weight than dollars. Do we know how to include the human aspects when evaluating the risks of pesticides and nuclear reactors? How about airplanes and cruise ships? Would our method have saved any lives on the *Titanic*? As we can see, there are fundamental problems in this risk assessment business. Now we are ready for Quantitative Risk Assessment, QRA.

In the early days of tackling questions like this, about 1940 or so, many of these problems were avoided by a concept that seemed logical and very safe. The idea is straightforward enough. Suppose we wished to set a health standard for chocolate flavoring. Initially, there are two possibilities: The flavoring is absolutely safe, e.g., no matter how much we eat, except for maybe a sour stomach, it will not hurt us; or, there is some level of ingestion, the deadly threshold, at which the substance becomes toxic. To check this out we make some assumptions. First of all, we assume that animals, especially rats, react to chocolate flavoring in exactly the same way people do. We choose rats because we will probably kill a lot of them in our testing and not too many people will become too excited over dead rats. In addition, as laboratory animals go, rats are fairly inexpensive, they are small and don't take up a lot of space, and they are mammals, so maybe they will respond to our tests or treatments in a manner at least similar to that of people. Second, we assume for a safety factor, the deadly threshold for people may be ten times less than that of the rats and also that there may be a tenfold variation among people. So, if the rats do die from chocolate flavoring at level X, then ($10 \times 10 = 100$) the safe level for people is one one-hundredth ($\frac{1}{100}$) of X. To run the test all we have to do is expose a few rats to a limitless supply of chocolate flavoring, record how much they eat and when and if they die.

Let's review the rationale once more. It's very simple: For any poison, toxin, or what have you, there is a threshold below which the substance is harmless and above which it is deadly. All we have to do is test rats to find that threshold. The threshold came to be known as the "No Observable Effects Level" (NOEL). The NOEL method was popular for several years until someone suggested that maybe the whole idea was wrong. Instead, it was postulated that the threshold concept was incorrect and even less than the tiniest amount of toxic substances is bad and will be dangerous to somebody sooner or later. Exposure to more than the tiniest amount would simply bring doomsday sooner. This no-threshold idea was like the S-curve of the last lesson. At very small levels, the chances of death are small, but they are still there. Increasing the dose just pushes us a little further up the curve. This idea gained adherents and is still the underlying premise of QRA. As we might suspect, no one knows which premise is right, the no-threshold idea or the old NOEL concept. Chances are that sometimes one is right and sometimes the other, depending on the substance being tested. Of course neither premise would apply to a common item like table salt, a substance about which we know a great deal: We need a little to be healthy, too much may raise our blood pressure; and a whole lot, as in sea water, will kill us. This salt scenario has aspects of both the NOEL and QRA theories, with the added kicker that a little is not only harmless, but even essential. Thus, both ends of the salt intake spectrum are deadly. In addition, this same statement is true for some minerals, such as manganese, declared as "heavy metal toxins." This demonstrates that both the NOEL and QRA ideas are wrong at least some of the time. The thing to remember from all this is that the very foundation of QRA is unstable. What's more, the foundation is the best part; while it is merely unstable, the structural glue holding QRA together is spittle, as we shall see below. This is where the statistical tools take over.

At any rate, with the rejection of NOEL and the acceptance of the no-threshold (I'll call it the S-curve) concept, a whole new ballgame was started. This game has several parts, each of which generates its own feeling of sinking confidence. For starters, let's return to the rats and the chocolate flavor, but this time we're looking for cancer of the appendix. We need another assumption: a *small* number of rats with a *lot* of chocolate will tell us about a *lot* of people with a *small* amount of chocolate. This assumption, which is supposed to explain about lots of people with very small doses is a must. Note, however, that it is also another form of extrapolation born of necessity. Very long testing with large numbers of rats is simply too costly in terms of time and money. These extrapolations may or may not be true. For example, let's assume that we want to assess the risk associated with a tiny leak from a (chocolate?) reactor that lasts for 2.7 seconds and is thinly spread by the wind over all the Great Lakes, plus 190 counties and provinces, including four major cities. We will simplify this test by eliminating long-term effects. It is assumed that the chocolate-flavor component of the leak will decay to zero naturally within forty-nine hours so there can be no lingering contamination. You can see why this assumption is necessary and how insidious things like slow accumulations, as in lead poisoning, could confound all of this. How would we equate four pounds of lead in five rats over two weeks to a milligram of lead in a large family over six years? In addition, if cumulative or long-term contamination must indeed be considered, yet another group of experts is needed because that demands still another kind of study, e.g., how molecules in the air get into fish in the ocean, that get into nets, that get to fish markets, and so on. There is no need to belabor the details. Just remember that real studies encompassing QRA must include them.

Well, to wind up this first part of the new ballgame of no-threshold, in our tests, we found that 50 percent of the rats who jointly consumed eighty-seven boxes of Sarah's Samplers in five days had malignant appendices the following week. A phone call to Sarah and a little arithmetic let us calculate that the eighty-seven boxes was equivalent to fifteen ounces of pure chocolate flavor. Having done this you can see how "easy and obvious" it must be to extrapolate to the conclusion (to be a benchmark) that one part of chocolate flavor per 27 million parts by weight of a human body will cause cancer of the appendix in one person in 853,000 (probably). Naturally, this estimate involves the use of numerous statistics on body types, physical activity, overall diet, and so on.

So much for the first part of this QRA assessment task, the second one is *difficult.* This entails determining how many people, if any, in all of those 190 counties and provinces, will ingest enough chocolate flavor from that 2.7 second leak to effect a contamination of one part in 27 million. Well, there are a lot of variables here: wind speed and direction, the amount and frequency of precipitation, the likelihood that it is a holiday weekend when many people will be swimming and so partially protected from fallout. The list of variables could be quite long. The experts agree that they cannot know all of these things in advance and so they agree, with only a few dissenters, on a worst case scenario. (Here is the real truckload of subjectivity for this project.) The worst case is defined as this: A major event has occurred near the Great Lakes and everyone has visiting family from all over the country, the population is twice its normal size. It is a glorious spring day, everyone is outside wearing light clothing. The wind has uncannily distributed the leak material uniformly just about everywhere so everyone is exposed, but concentrations ten times normal are left in critical locations. (This last condition is a compromise between the contentious even distributor and clumpy fallout schools of thought.) A critical location is defined as anyplace that has at least one super-chocolate-sensitive person, identified in the report as an SCS subject. Of course, no one knows for sure that there are such things as SCS subjects, let alone how many: They are simply postulated for worst case purposes. Furthermore, since we don't know how many, or where they are, we will assume that anyone who may become ill as a result of the leak is, unfortunately, an SCS type. The fact that this chain-reaction type of reasoning makes the leak effectively ten times worse than it actually might be is of no concern. If anyone raises the point we can call it a safety margin allowing for unknowns.

A check with the designers of the reactor determines that, in the worst case, the pressure could be so great that the 2.7 second leak will release 4.8 grams of pure chocolate flavor into the atmosphere. Now that we have all the facts about this potential leak—who, living where, might be affected, and what the chances are that this flavoring will destroy an appendix—we will take the analysts' word for the estimated fact that there is one chance in 907,000,000,000,000,001 that your Uncle Hiram will be a victim.

The reactor leaking chocolate flavor is a simplified case used for illustration and to get the basic procedure in focus. In a real-life study, a real-life leak, from a reactor, incinerator, or anything of interest, will have not only chocolate flavoring, but dozens of other varieties of suspected toxins, carcinogens, and generally dangerous substances. Each of these must have its own study with rats, or maybe with rhesus monkeys in case rats aren't affected. Each may be distributed in different ways, some on the wind, some seeping through the soil, some drawn high into the atmosphere to fall in rain (maybe in Washington, D.C.); and each has its own half-life (the time for its deadly properties to become half as potent), some of them 10,000 years. The last part of this new ball game is to combine the risks of all of these nasty things into a grand chance of death. And the answer is: any number from zero to one that you care to put here.

Why so cavalier about the number? First of all, numbers coming out of these studies are about as firm and final as a morning mist in the Sahara. Current dramatic evidence of this is the turn-around on the permissible levels of the toxic substance dioxin: a decade ago it was thought that exposure to one part in a billion was dangerous (a view that brought about the federally funded abandonment of an entire town). It's twenty times that amount today. Even though there is a fair amount of sophisticated mathematics used in the derivation of these numbers, every step of the game is plagued by subjective elements: If eighty-seven boxes of chocolates killed the rats in one week, how sick will a 100-pound woman be after eating four boxes of Sarah's Samplers? What if she drinks a lot of water? Will it make a difference whether she is twenty or seventy-four years old? Obviously we cannot test women to answer these questions, that's why we use the rats in the first place. In fact, we often don't even know if chocolate flavoring will have any adverse effects on humans even if *all* the rats drop dead. Remember our starting assumptions about the rats being mammals available for tests? Furthermore, when we mentioned the rhesus monkeys (in case the rats aren't affected), we were not kidding. If it were found that the chocolate flavoring did not affect the rats, then a search for some animal that is susceptible (even one less humanlike than a rat) would be undertaken. If nothing dies, how can we study and report on the chances of death? Also, wouldn't it be better to make an error and call something dangerous than to err in the other direction?

Now we have some idea of what QRA is and what it is supposed to do. The problems mentioned here are only a few of those plaguing both the concept and its implementation. The experts using it to generate numbers even disagree as to how the studies should be conducted, as well as what the results mean. However, at least there are folks working on the problem.

Having brought up the idea of human values in risk evaluation, this is a good place to again illustrate how meaningless risk assessments are, even real ones, let alone those based upon odds ratios. There is an interesting kind of public reaction test we can visualize.

Though apparently harmless, it could never actually be conducted for reasons that will be obvious. Suppose that the Federal Aviation Administration, feeling very safety conscious, required that airlines print on each ticket, and post at the gate, the chance that your flight will crash: "Flight 942, Destination Las Vegas, Chance of Crash: 0.000015." Or, after protesting, the airlines won the option of posting: "Flight 942, Destination Las Vegas, Chance of Getting There: 0.999985." Well, you know how soon the unemployment lines would fill up with pilots and stewardesses. Uninformed potential passengers, seeing such information on the flight status board would dash for their cars to drive to Las Vegas, with the chance of crash only 0.0015. It's much safer this way, 0.0015 is smaller than 0.000015 by two whole digits, don't try to kid me! Now, if you think you really believe in risk numbers, tell me, how small would the number posted at the gate have to be before you would board the plane?

If that very same number, 0.000015, was the chance of winning a multimillion-dollar lottery, the exact reverse would happen of course. More tickets than ever would be sold (if the prize stayed the same). And of course we know the reasons for this reversed response.

Attempts to formalize these dramatic and varied reactions to the same values of chance have developed into a mathematical methodology known as *decision theory*. This is another esoteric subject, very likely applicable to the risk evaluation problems we've been discussing but so far apparently not yet used by those workers. It is interesting and, on the surface, shows a lot of promise. In the final analysis it is just another way of pushing numbers around, once more leaving the final choice up to us and our gut feelings. Its potential advantage is its formal and perhaps rational way of reducing a great many risk numbers down to a choice between two. And, if we really follow the rules, the "choice" is made before the arithmetic starts, so the answer is arithmetic. (Even so, your gut feeling may get hung up on one of the minor inputs.)

To get a handle on decision theory, let's go back to that common number appearing on the airline ticket and the lottery advertisement, 0.000015. I believe we agreed that it would chase people away from the aircraft but attract them to the lottery. The reasons for the reversed response are the ones theoreticians attempt to quantify for use in their decision-making models.

Looking again to the gamblers, which is where formal probability originated, the theorists discovered that there is something (often subjective) known as a "payoff" and it seems that gamblers are more willing to put money down if the payoff is big, with seemingly little regard for the chances of winning—what they call the "odds." The theorists wondered if this payoff was the clue to understanding why people gamble. They could translate the payoff into numbers, which allowed for negative payoff (a gambler's losses) as well as positive. This payoff idea is also more flexible than plain chances, the numbers can be any size, not restricted to values between zero and one. Obviously the attractive payoff is usually tempered in some way by increasingly poorer odds or chances. The greater the payoff the lower the odds of winning. For starters then let's just add the two, payoff plus odds, the thought being that, on average, larger payoffs balance smaller odds; maybe we can generate some sort of overall worth number—a barometer of why people gamble.

Consider the choice of playing any one of three $1 slot machines. The first machine pays off $10, the second pays off $100, and the third pays off $1,000 for the winning combination. Well, since any of them cost only one dollar, isn't it obvious to play the big one—the $1,000 machine? Maybe, but we're in a casino, and let's postulate that all games are adjusted so that, in the long run, the house will keep 10 percent of all money played. In other words, for the house to pay $90, it must take in $100. If this is true then the chances of winning on the three machines must be: 9 in 100, 9 in 1,000, and 9 in 10,000, respectively. The chances of losing are the complements of those numbers, 91 in 100, 991 in 1,000, and 9,991 in 10,000.

To evaluate the three slots, the idea is to consider a single $1 play. Multiply the possible winnings by the chance of winning, e.g., $\$10 \times (9/100) = 0.9$, and the possible negative loss by the chance of losing, e.g., $\$1 \times -(91/100) = -0.91$. These two (worth) values are then added to get a net value. Doing this for the three machines, 1, 2, 3, we have the following:

SLOT	**1**	**2**	**3**
Win	–0.9	0.9	0.9
Lose	–0.91	–0.991	–0.9991
Net	**–0.01**	**–0.091**	**–0.0991**

The decision to play is made by choosing the best of the "Net" numbers. Of course they are all negative because the odds are tilted in favor of the house and we really ought to get out of the casino.

But, in the true gambling spirit, enlightened now by this set of "weighted" losses, we play machine number 1 because it has the smallest loss value (it's the least negative).

To get a better feel for how such numbers can change and what relative meaning, as well as values, they can assume, let's look at the next three slot machines in line. These have the same payoff values, $10, $100, and $1,000. They are more intriguing however; they appeal to the big spender because they cost $1, $10, and $100 to play. Now you just know that the odds must be better here.

Assuming the same rule about the long-term house take, the chances of winning on these three machines work out to be all the same, 9 in 100, and for losing, 91 in 100. Here are the tables for these three just as we did above:

SLOT	**1**	**2**	**3**
Win	0.9	9	90
Lose	–0.91	–9.1	–91
Net	**–0.01**	**–0.1**	**– 1.0**

Again we play the first machine to take advantage of its smaller loss value. Notice however, that while the net loss value for the first machine is the same in both cases, –0.01, the weighted penalty is more severe for machines 2 and 3 in the second set than for 2 and 3 in the first set. Of course that's what we would expect because of the much bigger cost to play, $10 or $100 versus $1. When we lose here, we lose big. The two net values comparing the number 3 machines, –0.0991 and –1.0 are not nearly as emotive as actually losing $1 compared to losing $100, but that's just a reflection of the aloofness of mathematics. We have to develop a trust in what the numbers are telling us, even if the differences seem small.

The first strategy we chose to impose on the gambler making his choice of slots is one that will cause him to lose the least amount of money in case luck is against him. At the same time however, he will win the least amount of money if his stars are right. This conservative strategy is known as "minimax," for *mini*mizing the *max*imum loss. There are other strategies, involving different weighting techniques, that we may choose to use. In case we did not like the arbitrary choice of minimax in the scenario above, one of the many others may suit us. For example, one of them calls for maximizing the expectation. Another wants us to maximize the negative regret.

With names like that, need I say that statisticians invented them? What all these techniques do is reduce the number of numbers we have to look at and then apply a rule to pick one of them, just as we did with the slots. I hasten to point out what you already suspect: Applying different strategies to the same problem may very well cause us to make different choices even if we follow the rules. As I said earlier, in the final analysis these are just different ways of pushing numbers around, once more leaving the final choice up to us and our gut feelings. In these cases, our choice is from a group of strategies. In the end, in spite of the number crunching and our, let us say, up-front decision to be conservative, when we get in front of the slots our gut feelings may force us to drop a $100 chit into that number 3 machine anyway.

I'm confident at this point that you can see how these decision strategies might be incorporated into the risk evaluation game.

A simple case would be an application in hazardous waste cleanup. If we assume that the relative dangers of various contaminants are really known, at least to the same degree of uncertainty, then we can assign the risks to them—lead: 0.05; mercury: 0.08; etc. These numbers must be multiplied or "weighted" by the amount of each of them in the dump considered. If, for example, the dump had only a trace of mercury but a lot of lead, the lead would be of more concern so it would receive a bigger weight even though the mercury risk is larger. Then we add up all the weighted toxins to get a "badness" index for that dump. This index is a penalty associated with the dump, much like the cost of losing at the slot machine. But we are not through yet, we need yet another weight. If the dump were found where a lot of people get close to it (Love Canal is a good example), it would be more important than if it were on an uninhabited remote island never visited by people. So, after deliberating and consulting, each dump receives a number, say from 0.0 for the island to 1.0 for Love Canal, for its location index. Multiply the badness index by the location index and we get an overall penalty rating for each dump.

Now it is time to look at the economics, the estimated cost to clean up each of them. Don't get alarmed and say that if lives are at stake, cost is of no concern. We want to address that very issue. We know that cleanup funds are short so we look for the biggest payoff, as we might have with the slot machines. If we built the rating correctly, then the number of people potentially at risk is plowed into it. So, if we divide the rating for each by its cost, we have a number representing people removed from risk per dollar spent. We might call it the life retrieval index. That would make everyone want to maximize it, and perhaps accept the (resulting) priorities for cleaning up dumps.

The hazardous waste cleanup can be approached with a rather straightforward application of the concepts of decision strategies. In other situations there may be some very real problems to overcome. Recall that for the slot machines, the only subjective part was deciding which strategy you were going to use, the (odds and cost) numbers were objective. This is not always the case. Suppose we really wanted to evaluate the benefits of saving a lot of lives by simply eliminating automobiles. Besides evaluating real dollar costs of that, we must come up with numbers expressing our relative feelings about less smog, fewer deaths, walking to work, and shipping by horse and wagon, or whatever alternatives are offered. Each of these is subjective and feelings vary from person to person. One approach to generating a set of such numbers is a simple count of yes and no votes on each item in the list. Even then, there are those who would scream about tyranny of the majority. The people trying to work with QRA are continually faced with this sort of dilemma. It is a major reason that we have not seen or heard of more results from that effort.

Lesson 6

The Real McCoy

I must admit that in the last lesson perhaps I got a little carried away with my description of QRA and what it doesn't do. After all, it is still in the formative stages, I have no idea how many, if any, regulations are currently based on a QRA analysis. However, when they come, you will know all about them.

For now, let's get back to some real-world techniques and criteria. Caution, this lesson is tough, I'll have to get more technical than ever. As with most difficult things, the rewards are large. Here is where we discover why studies are forever contradicting each other.

In lesson 4 we introduced the S-curve or the so-called logistic function. I also talked about risk ratios and indicated that the two were somehow related, but left the relation in a fuzzy state. That was intentional. The S-curve is a handy way to visualize how things can creep up on us or suddenly seem to overpower everything. In that sense it quantifies a lot of human, and other, natural activity. In the world of practical statistics it has a more clearly defined role, at least as used by researchers in the health and medical fields. The S-curve is representative of one of three broad classes of techniques used in these areas and it sits in the middle of the three, both historically and with regard to popularity. In my opinion it is also in the middle in terms of believability of results, but that is, of course, subjective.

These three classes will be referred to as table, logistic, and hazard.[1] These are my shortened names for contingency table, logistic model, and proportional hazards. There are actually a number of different tools within each class, but within a given class, the tools are similar. The classes were developed and used in the order listed, and of course, each is more complex than its predecessor. All three are still in use, and I suspect every research group has its favorite; the published papers in technical journals indicate that. Furthermore, there are well-defined types of experiments, each of which is thought to be best suited to one of the three classes.

The table method, the oldest and most straightforward, had its origins in biological-agricultural experiments, for example, to make comparisons between fertilizers or seed types. Formally, the table method lets us compare things by writing down our experimental results in a table with two rows and two columns. Each row has information about one of the two things being compared, say, seeds. Each column lists what happened to the two things under different conditions, more or less water, fertilizer, etc. Then we make a hypothesis. There's no need to dream one up, it is always the same: "There is no *association* between the two things I'm comparing." There is another official word for our vocabulary—association—which is a fancy way of saying there isn't any difference between the things described in the two rows. Of course, this is the null hypothesis, and in most cases we hope it is wrong.

Suppose you were in a craps game and had cause to suspect that not all was on the up and up. In particular, you suspect that a certain pair of dice is loaded. So (in this fantasy world you can do this) you take the suspect pair of dice, and a known good pair, maybe from your pocket, and throw each of them until you get a total of one hundred sevens and twos (snake eyes) from each. If the suspect pair is loaded, the numbers of sevens and snake eyes should be different than for the good pair.

Now you put your results, showing how many twos and sevens you threw, in a table like this:

	Twos	**Sevens**	**Total**
Good	3	97	100
Suspect	5	95	100
Total	8	192	200

Table 6.1: Contingency Table for Throws of the Dice

Which shows that you had three **twos** and ninety-seven **sevens** from the good pair but five **twos** and ninety-five **sevens** from the suspect pair. On the right and bottom margins are the total number of tosses and the total number of twos and sevens, for each row and column, and then the grand total, two hundred. Now, would you believe these totals are named "marginal totals"?

Well, right off the bat, you're suspicions are confirmed, the numbers are different, your dice give more sevens and fewer snake eyes than the suspect pair. No, no, no, this is statistics, you haven't made any tests, so you can't make those brash statements yet. What is needed here is some measure to tell us whether we should believe that difference (maybe it was only a fluke). We need something akin to either the expected error or the confidence or both that we calculated for survey results. In this case we will calculate a *significance level* at which to *accept* or *reject* a belief in the difference. This is the very same measure that researchers use to decide if their newly found risk is believable. It is how they decide to drink the coffee or not.* What we have to do is estimate the chances of twos and sevens, from your data in the table, and compare those with what you actually got. Sound good? Okay. To make the estimates, use the marginal totals.

On the right, the totals say that there were one hundred of both kinds of tosses with each pair so, if there is no difference between the sets of dice (our hypothesis) then the chances of twos or sevens should be the same. This means that the total number of twos, eight, should be evenly divided between the two rows, just as the total of two hundred was evenly divided between them, and similarly for the 192 sevens. That division is obvious here with the nice numbers, but the probabilities are really calculated as $^{100}/_{200}$ for each case. That is, divide each righthand marginal total by the grand total to get the chances associated with each row.

Now, take the total number of twos, eight of them, and multiply by the chance for each case ($^1/_2$). We get the *expected* number of twos, four, for each pair. Similarly, $^1/_2$ times 192 gives the expected number of sevens, ninety-six for each case. Now look at the table, the actual number of each kind that you threw was never what is expected. Now what?

Without getting into all of the arithmetic, we make a significance test that involves taking the differences between the numbers in the table, the observations, and the expected numbers that we

*A little later we will see how believable their significance levels are.

just calculated. Then we square and divide and so on. At last we get a single number and look it up in one of those probability tables we mentioned in lesson 2. In this case it must be one called a chi-square table of numbers, or deviates if you wish. The table in this case yields a significance value of 0.86. This is awful. Remember, the significance number must be very insignificant because it is the chance of being wrong if we believe the difference (reject the hypothesis of no difference). Such a large number would support the view that there is no difference in the dice. So, instead of rejecting the null hypothesis, the data tend to support it. As a matter of interest, it turns out that in order for the significance to be 0.05 or 5 percent, a borderline kind of value, you would have had to roll ten deuces and ninety sevens with the suspect pair. For a 1 percent significance, the number of twos would have had to be thirteen.

Well, if you were the big loser in the craps game, that little exercise probably did not impress you. Go right on believing that those dice were loaded, no matter what the statistics said. The point here is that the table method is generally conservative (subjectively that is) about rejecting that "null" hypothesis. It is not foolproof however. Many good textbooks will give examples of how the contingency table can give wrong answers if there are more than two variables involved. A common one has to do with the relations between drinking, smoking, and heart disease. If we are not careful, it is easy to put the blame for heart trouble on drinking rather than smoking, where we all know it belongs, don't we?

Of course, the simple 2×2 (two by two) table, as it is called, can be expanded to allow for practically any number of variables. In addition, some fancier arithmetic is introduced, allowing us to extract more information. And, of course, new names are used. One sees reference to things like anovar, latin squares, block design, etc.,[2] but they all revolve around the basic table concept. In the dice example, we could have recorded the totals for all eleven possible numbers on the dice, two through twelve, and made a 2×11 contingency table. The only real difference is that more arithmetic is then involved. So, the computer works for another fraction of a second. The advantage is that the decision making would be a little sharper. That is, because there would be more evidence to test on, it just might "prove" your suspicions without having to throw thirteen deuces.

The table methods have the attractive feature of being more intuitive than the others. In the table above it is fairly easy to see that, while the number of twos for the different dice are not the same, they are close. Also, estimating the chances with the marginal totals is fairly obvious. We could see instantly that it was a 50-50 situation. I had to point out the formal calculation of the numbers ($^{100}/_{200}$ = 0.5 in this case). Of course, with a large table, like the 2×11, significant items most likely would not jump out at us as quickly, but we would have a feel for what the arithmetic was doing. For the other two classes of techniques we shall see that things become a little more abstract.

At this point we might ask why, if the table method is so straightforward, is there any need for those abstract techniques? Well, real-world problems can get messy. Consider the heart disease, drinking, smoking case alluded to above. We start out by asking the subjects whether they smoke (yes or no) and whether they drink (yes or no), supposedly we know the condition of their heart. The contingency table is still pretty simple. It is really just two of the 2×2 tables side by side, and might look like this:*

	Smoker Heart Disease		**Nonsmoker Heart Disease**		**Total**
	Yes	No	Yes	No	
Drinker	55	23	4	8	90
Nondrinker	5	2	11	21	39
Total	60	25	15	29	129

Table 6.2: Contingency Table for Effects of Drinking on Smokers and Nonsmokers with Regard to Heart Disease

Then we would proceed as above with all the arithmetic for each of them with the results that the relative odds for heart disease for smokers who drink or not, the left 2×2, is 0.96, and for nonsmokers who drink or not, the right 2×2, is 0.95, essentially the same result. That is, drinking makes no difference in one's risk for heart disease whether one is a smoker or not. Remember, it's the two rows that are being compared, not the sets of columns. The wrong answer mentioned above comes about if we ignore smoking and just look at drinking by combining the two 2×2 tables into one. This is what would have happened if we were not smart enough to include smoking in the tests. So, add up the total cases of heart disease for the smokers and nonsmokers to eliminate that as a variable:

*These are hypothetical data. Results are only illustrative.

	Heart Disease		
Drinker	Yes	No	Total
Yes	59	31	90
No	16	23	39
Total	75	54	129

Table 6.3: Contingency Table for Drinking Only

Now the relative odds for the drinker versus the nondrinker for heart disease is 2.7 to 1. The significance for either of these tables is acceptable, about 1 percent for this last one and (wow!) 0.0000000003 percent for the longer table. This last result does indicate that the long table is to be preferred. Sound's good, doesn't it? But in how many studies is it known that all the proper variables were included? This is our first example, with numbers, of how study results go bad because of ignorance of other factors. If any critical (often unknown) variable is left out, the results are simply wrong.

When we consider two or more variables, drinking and smoking in this example, it is often the case that one or more may be confounders for another. A confounder is a variable that is a risk factor for the disease under study and is also related in some way to another, nonrisk variable. For our example it would be customary to think of smoking as the confounder for the study of the effects of alcohol. Smoking and alcohol use are related simply because people who drink also tend to smoke. However, as you are no doubt aware, there is still controversy over the good or bad effects of alcohol with regard to heart disease. If alcohol is bad for the heart, then the two are intertwined in some way and extraction of the true risk for each is very difficult. The confounder is whichever variable with which we are not immediately concerned. The point is, not only for table-type analyses but for any type, if we don't have every possible confounder accounted for and then handled properly through independent tests or some other method of separation, we will get wrong results.

As if that's not bad enough, in principle we are doing the tests to look for (not prove) possible causes of some state of affairs. So, we want to include lots of possible confounders and get a whole host of different answers, depending upon how many we include in each of the many tables. How do we then decide which of the many results is, or are, correct? Flip a coin? Pick the ones we like? The fact is, we should not pick any. The only legitimate reason for making these tests is to find out which of the confounders may be of interest. Each test is only a guide to future, or maybe a buttressing of past, research. Statistical tests do not constitute research. They will never explain anything. Mendel's laws of heredity did not arise from any statistical analysis, as many texts would have us believe. They followed from the manner in which he conducted his experiments.

To continue with why contingency tables are not enough, think of an even more complex case. Instead of just drinking, yes-no, and smoking, yes-no, we might wonder whether the amount of each makes a difference. We could break up smoking into none, half pack or less, half to one pack, one to two packs, and more than two packs per day. For drinking we might try none, one a week, two or three a week, one a day, and so on. Now, with only two variables, the contingency table is at least 2×9 and, unlike the eleven dice numbers, where we were interested only in the full-table analysis, here we want to know, for all possible combinations, is drinking important or not when considering smoking and heart disease? If we now toss in a few more variables such as fat intake, use of contraceptives, hormones, and coffee, with each at several levels, the tables get completely unwieldy. The computer could handle it, but in general, all those results would simply overwhelm us; we couldn't begin to understand them.

Here comes the logistic S-curve to the rescue.

It is widely known that researchers, scientists, mathematicians, and so forth, make models of relationships between variables (actions, qualities, etc.) In spite of the jokes about their sand boxes, most of them don't sit around gluing plastic replicas from toystores.

Mathematical modeling may sound mystic but the ideas are really quite simple. Recall that the picture of the S-curve in lesson 4, repeated here as figure 6.1, had that ambiguous arrow along the bottom. We glossed over it with some words about representing increasing risks. Suppose I said that each little step along that arrow represented another swig of booze, and the height of the curve at each swig indicates how drunk a person is instead of the chance of being afflicted with something or other. Or, we might even think of it as the chance of being blind drunk, on a scale from zero to one.

There is more reality here than we think. Most people can stand a sip of wine with no effects. That is represented by a point on the far left of the arrow where the S-curve is hugging the bottom of the picture, near zero. As the number of sips increases, the effects begin to show as the curve rises above zero and by the time we reach the middle of the arrow the drinks are piling up on us in a hurry. The

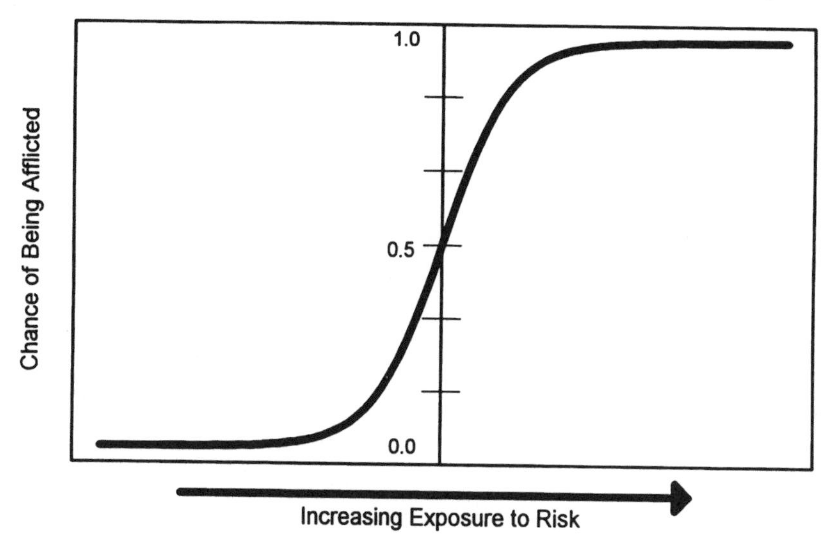

Figure 6.1: The Logistic Curve

center of the picture might be the "whoopee" point. If we continue sipping or moving along the arrow, we finally reach a point of total intoxication, the curve flattens out, and probably, so do we. The S-curve can describe the effects of alcohol and these can be used as a mathematical model of drinking. To use the model, the arrow gets calibrated in number of sips, or ounces of alcohol (say, perhaps, per hour), that's what we call "fitting" the model; then for any stated number of sips, the model tells us how drunk someone is. Simple, isn't it? Of course, the calibration or fitting part is crucial. A great many people must be tested with all the various amounts of alcohol and then how drunk they are measured. That is really what the studies are all about, curve fitting, pure and simple. If we are good at it, or lucky, we can fit the curve by using a fairly small number of drinkers and then apply it to everyone in the country.

"Fairly small" here is of course a relative term and a little different than in the case of surveys. For surveys, we could estimate as part of the planning how many people we would have to interview for a certain maximum error. That is not possible here. We don't know how good the result is, i.e., how well we fit the model, until all the results are in. We discover how well we did by the size of the significance level. It is calculated along with the risks we estimate from the model. And, as we will see later, the total number of people in the survey does not directly affect the result. It is the number of actual "cases"' that is important. Which leads to another word of caution: Don't be impressed by studies that advertise hundreds of thousands of subjects; there are typically only a few hundred cases.

Now, if we can replace ounces of alcohol with arbitrary mixtures of smoking, fat, hormones, lifestyle, and other helpful variables, we can predict heart attacks or cancer or just about anything with an S-curve; all we have to do is calibrate it. This is where it gets interesting. But first, a few words about how we can include arbitrary mixtures of all these items. Arbitrary mixtures can be included easily because of the equation chosen to represent the S-curve.

Many mathematical equations can generate an S-shaped curve. The cleverness comes in picking the right one. In the introduction I promised that there would be no mathematics beyond simple arithmetic and there is really no need to write one for the logistic curve but I will show an example in case there is some interest. If you skip over, you won't lose anything. Suffice it to say that it is simple, and such that every time I add another variable in my study, I can simply toss in one more parameter (another letter) in the equation and, although it may get longer it never gets unwieldy.

The example:

$$\text{Chance of disease} = 1/[1 + \exp(B + RV1 + RV2)].$$

B, R, and V are numbers that the computer generates from the data we give it—the fit to the curve. "Exp" is a notation meaning "exponential function." B represents the "base" chance of getting the disease; no known exposures to bad (or good) variables. RV1 ($R \times V1$) represents **R**isk associated with **V**ariable 1. RV2 has the same meaning, but for variable 2. The number for each R is not the risk exactly, but the risk can be calculated from it. The terms RV1, RV2, etc. can be thought of as adding to or subtracting from (in the case of good variables like exercise for heart disease) the overall chance of the disease. In principle then, anyone could find all the RV terms applicable to him- of herself and estimate the chance of contracting the disease, i.e., how far up the S-curve he or she is.

When I want to look at the effects of only one variable, I can simply set all the others equal to zero and still get the "correct" answer, the one that applies to that special case. Setting one of the variables equal to zero is, in the model, roughly equivalent to saying "don't include any drinkers" in the calibration or curve fitting. On the other hand, if I want to know about the combined effects of two variables, I set all but those two equal to zero, and so on. In addition, I can find that relative danger of two conditions, the odds ratio, of any variable compared to any other or to B, the base, in a very easy way. Because the equation gives the estimated probability of getting sick, or dying, I can arrange it to calculate odds and then just zero out everything but smoking. This is almost too good to be true. The mathematics is impeccable, but the users had better be careful. Remember, they have to calibrate the whole thing and how reliable that calibration is depends upon how good the input data were as well as what significance level they like.

Well now, this logistic model appears to be the answer to everything, but is it? In the area of medical studies two fundamentally different approaches are taken. One is to collect a group of healthy subjects and follow them all as closely as possible for some period of time to see who succumbs to what and, hopefully, for what reason. These forward-looking types are called *prospective* or *cohort* studies. The second approach is to collect a group of sick or dead subjects, referred to as cases, and compare them to a carefully selected similar group of subjects who do not have the particular problem. The latter is of course the control group. These backward-looking types are called *retrospective* or *case-control* studies.

In principle, two of the three tool types, table and logistic, can be used to analyze the results of either type of study. However, in a cohort study, because we are tracking the subjects, information about the time of occurrence of disease or death (or events as they are sometimes called) is available. It's possible that these time-of-occurrence data could be important. Suppose that we were testing an antisomething drug to ward off a killer bug and we gave it to half of a group of potential patients. If we then waited perhaps three or four years and the entire group was dead from that bug, what would we know? In an extreme case, perhaps the nontreated patients all died within the first six months and the remaining succumbed gradually over the following three and a half years. We would suspect that the drug was helpful, but a table (see example in table 6.4) or logistic analysis does not recognize time and would come up with a no-benefit result simply because everyone was dead. These models are indifferent to the sequence of events.

Subjects	Alive	Dead	Total
Treated	0	150	150
Not Treated	0	125	125
Total	0	275	275

Table 6.4: Problems with a Table Approach, Time Not Considered, So Everyone Is Dead

One other annoying thing that happens in a cohort study is that some of the subjects drop out along the way for a host of reasons, often unrelated to the study. They might move out of town, die accidentally, or choose not to continue with the study. On the other hand, we may find opportunity to add new subjects to our group. So, for various reasons, the sample size keeps changing on us and we are losing information. People who drop out are referred to as having been censored. And, at the close of the study period, everyone left alive is automatically censored, but it is only the intermediate censorees or added new subjects that pose problems by messing up our sample size.

So, what we need is a way of using time information, allowing additions, and retaining some of the data associated with the prematurely censored folk. These are the reasons for the *hazard* model.

The hazard model has all the same useful properties that the logistic model possesses as well as being able to take into account event timing in its results. What it is trying to measure is however, a little more subtle than the logistic chance. The hazard model is close to being an estimate of the chance of an event (dying) at a relative point in time. Technically, it is often referred to as the instantaneous failure rate.

The concept is simple enough. Suppose we started out with five people, just to keep our numbers small. Four continued all through the study period, twelve months. One dropped out after three months and after six months we picked up another volunteer. For now, don't worry about who may have taken ill at what time.

All of this can be summarized as:

- Four twelve-month intervals from the four who persisted.
- One three-month interval for the dropout and,
- One six-month interval for the added subject.

The hazard model is made to be time or interval sensitive and uses times of survival or disease-free times (under various treatments), to make its estimates. So far everyone here is a survivor and the survivor times are those just listed. Now suppose some of the subjects became ill along the way, a possibility for any of them except the dropout who was censored while still healthy. While the four long-term folk could have become ill at times up to twelve months, the subject added is limited to six months to experience an event. Now, with illness occurring, some of those intervals above will be shortened and the reason will be, of course, that an event was recorded and that subject is now nonsurviving. The computer then can compare survivor times for event and nonevent subjects to make the risk estimates. There is a fine point about dropouts and full term survivors that is rather subtle but which will be easier to explain a little later, after we have learned more about the hazard model.

A related kind of model that is often associated with the hazard type is easier to interpret. We can talk about, and draw, a *survivor function.* This is simply a line resembling a poorly constructed stair case. At time zero, one starts at the top of the stairs with 100 percent of the subjects. As time progresses, events happen, and one proceeds down the stairs. Each step is at a level representing the percentage (or fraction) of subjects left. We'll look at some of these survivor curves in the next lesson. Sometimes it is easy to see and understand the effects of timing of certain variables being compared by looking at survivor function plots.

So, briefly, there we have the reasons for and the essential differences among the three types of models—table, logistic, and hazard—in common use. The logistic model has flexibility compared to the table, and the hazard takes time or, more properly, the sequence of events, into account. All three have their strong points and their shortcomings. As I said, in general, I think the logistic model is more believable, simply because it is more versatile than the table variety and the hazard is so flexible that it is also extremely touchy. As we will see, seemingly minor changes in the data can create massive changes in the results. To be used effectively, the data base underlying a hazard model must be impeccable in every regard.

It should come as no surprise that if we apply the three models to the same data (sometimes this is possible), we will, in general, get three different (often very different) answers. But, as I said, it appears that researchers have their own favorites. For a first look, suppose we take the smoking, drinking, and heart disease data from above and randomly assign a time or a sequence to each of its 129 subjects in order to make use of the hazard model capability. To alleviate fears of accidental bias, I used a random sample of times to assign slots to the 129 people with the intent of thoroughly stirring them up. In any case, applying the three methods to the resulting set of data, we calculate two sets of relative risk based on true risk ratio or chance, and two based on the odds ratio.

Before continuing, this is a good place to refresh our memory on chance versus odds, and point out why chance cannot be used in some studies. Recall that chance is the probability that something will happen. We estimate it by dividing the number of winners by the total number of players, and so our figure is always between zero and one. Calculating odds, the gamblers' choice, is done by dividing the number of winners by the number of losers, and may well vary all the way from zero to infinity.

Remember that if we do a case-control study as described earlier, we pick a sample of sick folk, usually just taking all that we can find. Often it may be everyone admitted to a hospital with a certain type of problem. These people become the cases. Now for controls, we need a representative sample of everyone who could ever come down with that same problem. Recall our discussions in lesson 1 about samples. At first glance, this means that the whole world should be represented. Well, if we work at the hospital it's convenient to pick controls from patients who are admitted for reasons not related to the problem that interests us. These people are not representatives of the whole world or even the fraction of it that may be subject to our problem of interest. Technically, we don't know what fraction of the susceptible population has actually been chosen for the controls. This fraction is called the sampling fraction and enters into the calculation of relative risks. Obviously, if we don't have that required number, we cannot make the calculation. However, if we calculate relative odds instead of risks, the sampling fraction cancels out, so if we don't know what it is, who cares? The penalty for this benefit, as we noted in passing in lesson 2, is that we can never know the basic chance or underlying risk of coming down with the disease. We don't know if we're dealing with a problem as common as the cold or one more like being struck by a meteor. In any event, for case-control studies, we are stuck with relative odds.

On the other hand, if we do a cohort study, our initial selection of a sample of healthy people defines, for our study, the entire population of interest, and later on we know what fraction became ill or remained healthy. Of course our problem then becomes one of convincing our colleagues that our sample represents the whole country. Note that whether we must use relative odds instead of relative risk is determined by the type of study we make. To be fair, I must point out that a hazard model can produce relative risks, even in a case-control study, if it is done properly (but see the next lesson for a discussion of that topic). For that matter, so can a logistic model) although that was seldom done until sometime in the mid-1980s. It is not clear why that seems to be so.

Now that we have risk versus odds clearly in mind again, we can look at our smoker-drinker results from the three analysis techniques. They are shown in table 6.5. Relative risk values, RR here, are interpreted in the following way. An RR value near one means that the risk of developing something is no different under the test conditions than it is normally. That is, the odds are 1:1, even money. If the RR value is less than one, then the odds are improved, it is less likely that a person will come down with whatever the condition might be. And of course, if RR exceeds one the person is at increased peril. In table 6.5 true relative risks are labeled RR, the relative odds are labeled RR (Odds).

	Table RR (Odds)	**Logistic RR (Odds)**	**Logistic RR**	**Hazard RR**
Drinkers	0.96	0.95	0.97	1.23
Smokers	4.78	4.78	2.08	1.78

Table 6.5: Relative Risks of Developing Heart Trouble Calculated in Four Ways from the Same Data

Notice that the table method and logistic method, shown in the first two columns, give essentially equal results for the odds ratios for both drinkers and smokers. However, the relative risks, shown in the last two columns and estimated by the logistic and hazard methods, are in disagreement. There are no real surprises here in view of the arbitrary sequencing due to the assignment of random times in the smoking-drinking data to use the hazard model. This is simply illustrative. In general, because the hazard model considers the sequence of events, it will almost always give an answer different from that of the logistic or table methods. In this case it made drinking appear to be bad, an increased risk of about 23 percent (1.23) when the others indicated it was unimportant; numbers just under 1.0 indicate no effect.

However it deemed the risk of smoking to be noticeably less than what the logistic method found, 1.78 versus 2.08 for the relative risk.

The contradiction about the effect of drinking might be cause for a little concern, but the two numbers (1.78 and 2.08) for relative risks are close enough that we could easily write off that difference in a one-study report and await further results. But note, for informational reasons, that the differences here (in the hazard model) came about by the introduction of random times for events, even though the "thorough stirring" should not have changed anything. Just keep that point in mind for now.

Nevertheless, the results actually don't look too bad, in that the consensus of all three analyses are nearly consistent, even though the numbers differ somewhat. By any test, smoking increases the risk of heart disease and drinking does not. That is, we are willing to concede that the observed 20 percent increased risk from the hazard model just calls for more work and must not be taken seriously now. That's grist for another grant.

Of course, we should go on and compute confidence bounds and significance levels for the numbers in table 6.5 before broadcasting the results. For now, simply note the likenesses and differences we found by using the three different tests on (nearly) the same dataset.

If we had proceeded with the computations just mentioned, then according to standard statistics, we would be finished with these data. In review, we designed an experiment to examine relations among drinking, smoking, and heart disease. We very carefully picked a sample of subjects, measured them, watched them, recorded events, and analyzed the data in a number of ways. The results, along with all the detail of the sample, would be reported and then carefully filed away for possible future reference. It's important to note that the sample detail must be maintained—not surprisingly, it is exceedingly important to statisticians. They worry incessantly about sample integrity. That's because much of the theory underlying their methods depends heavily on the sample meeting a host of highly technical qualifications. The rules governing sample selection and data handling must be enforced to ensure that the qualifications are met, or nearly so anyway. Very often, reports go into great detail to explain and justify the elimination of certain data points or the mathematical acrobatics used to overcome some apparent difficulty in the sample.

For reasons related to this obsession with sample integrity, statisticians never seem to perform one function that scientists and even engineers do all the time. Statisticians never examine results for sensitivity to fluctuations in the input. The input in this case is of course the data, and we might wonder how small changes would affect the results. The procedures are properly called *sensitivity analyses,* although various practitioners may use other terms such as "determination of a design margin."

Take, for example, the drinking and smoking data: we might wonder how the results would change if we had fewer than 129 subjects. This very thought is statistical blasphemy of the highest order. Nevertheless we will do it. Simply keep repeating the analyses we did above, but diminish the sample size each time. Don't take it all the way down to one, for obvious reasons, but stop at some reasonable point. We can determine such a point by watching the behavior of our answers, as we shall see below. When we are finished we might make a plot showing if and how the results change as the sample size changes.

First, use the table approach and estimate the relative odds for the effects of smoking and drinking on heart disease. As a matter of convenience and a way to keep track of things, use the ordering of the sample that was introduced with the sequencing for the hazard model. The result is shown in figure 6.2. We decided to stop at sample size thirty-eight simply because the scatter, or changes in the results, were becoming very erratic at about that point—a behavior typical of small samples.

Notice that I drew a horizontal line on the upper plot at a value for relative risk of 1.0, and labeled it the "Null Risk Line." That is the division line between "good" things and "bad" things, those that protect against disease and those that abet it. Points below the line indicate that the variable has a less-than-one or beneficial effect on heart disease. Points above the line indicate that the variable increases the risk of disease. The odds in that case are greater than one.

The apparent affect of smoking is shown, for each sample size, as an open square on the plot. The estimated odds ratios range from around five to one to about fourteen to one. We will write these as 5:1 and 14:1 for convenience. Even though the odds move around a lot, at least they are consistent in their statement that smoking is bad. With drinking, the solid dots, the results are not as clear. Most of them are near the line, 1.0, indicating that drinking is of no consequence, until the sample gets down to about sixty subjects, where it suddenly looks as though alcohol is a problem for the heart. Here the odds jump from about 1:1 to around 2:1. Does this begin to tie in with the reports you have read or heard about alcohol and heart disease over the past several years? Any one of those points could have been the end point of somebody's study.

Now look at the lower plot of figure 6.2 where the odds are plotted in percent bad or good (plus or minus), to provide a more symmetrical view. If we assume that this example is typical, researchers who had fewer than about sixty-five subjects in their test would claim that drinking was bad; if more than ninety and less than 120, they would claim that drinking was beneficial, and for more than 120, they would claim that there was no effect at all. *Here is our first real example of why studies disagree. In this case, the smoking risk is consistent, but the drinking risk is good, bad, or indifferent depending on how many people you tested.*

Let's perform the same analysis with the logistic model. In this case we can make two plots, one for relative odds, to compare with the table results; and one for relative risk, to compare with hazard analyses. The upper plot on figure 6.3 is the former. It is immediately apparent that even though the results are in general agreement with those from the table analyses, those of the logistic model are somewhat smoother. Notice the scales for the odds axis. In this case, the odds don't jump around as much, nor do they reach the extremes of the last results. This is because the logistic model uses a curve fitting process for each point. It is a kind of averaging, which is intrinsically a smoothing operation. The table method is a series of single-point estimates, so it lacks that smoothing effect. If that's a little vague, think of it this way. Recall the picture of the statistician's brush and the physicist's pen of figure 4.3. When we look at all those points, it is rather easy to draw a line through them that we can call representative. Our eye is using information from all the points to judge average values as the line is drawn. When one more data point is added to the "cloud" of points, as long as it is not a real outlier or otherwise effects a notable change in direction, our eye will not give it undue weight because it is still influenced heavily by all the other points. The line our eye wants us to draw will not move a great deal. Models that use curve-fitting procedures are doing exactly that. They consider where the new point falls with regard to all the others, as well as the numeric value of the point, and fit accordingly. Table-type models, on the other hand, respond only to the numeric value. They do not benefit from relative position information.

While it is a little difficult to tell from the two figures, their respective peaks and valleys appear to occur at roughly the same sample sizes. The variation in relative odds for smoking now runs

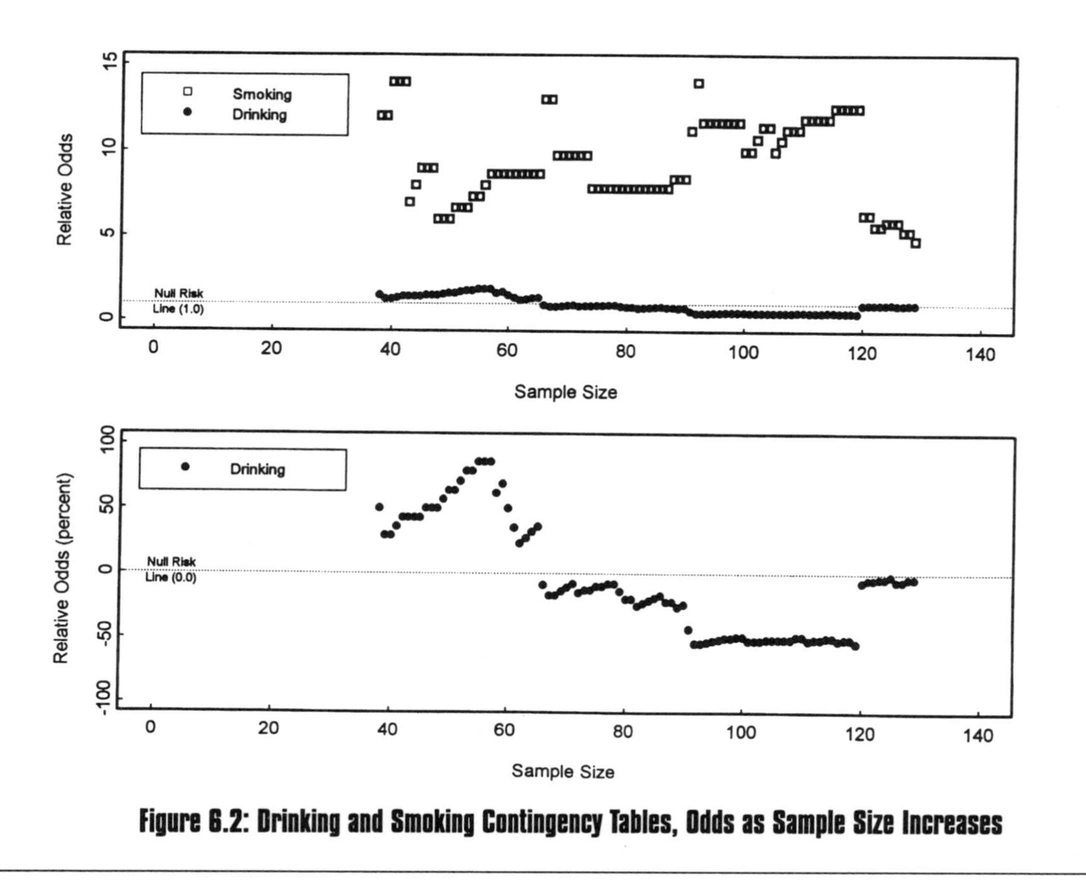

Figure 6.2: Drinking and Smoking Contingency Tables, Odds as Sample Size Increases

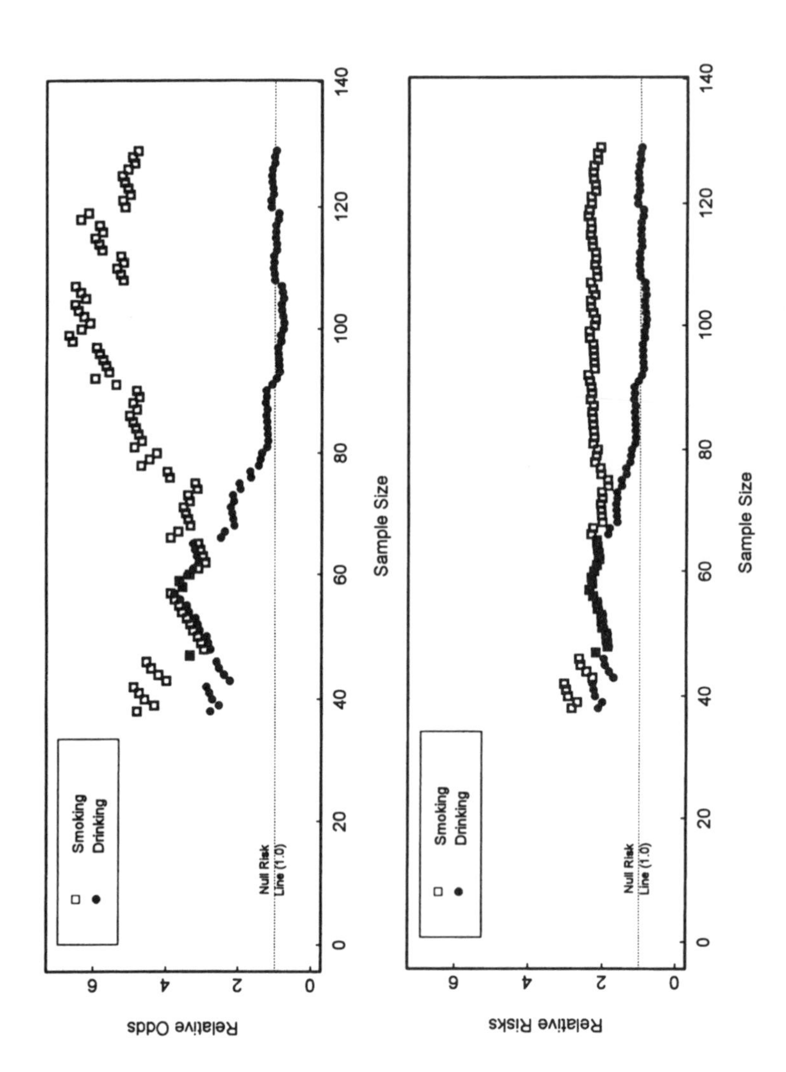

Figure 6.3: Drinking and Smoking, Logistic Odds and Risks As Sample Size Increases

from about 3:1 to near 7:1 instead of 14:1 as before. The range compression, or reduction in variation, is more evident in the numbers than by looking at the pictures.

Next, on the lower plot of figure 6.3, we have the relative risk values from the logistic model. Remember that in general, risk numbers are smaller than odds numbers. The patterns are the same, however, except that maximum estimated drinking risk now exceeds the minimum estimated smoking risk. The height of the upper black dot near 40 is actually slightly greater than that of the last two open squares, near 130. They overlap, showing how some reports could say that the risk of heart disease from drinking is greater than what others would report for smoking. The trends in risk with increasing sample size are the same as before, generally getting smaller with larger samples. We have a range of relative risks for smokers of about 2:1 to 3:1, and the drinking risk rises from 0.8:1 to over 2:1. So, depending on what sample size we happened to have, we would claim that drinking reduces our risk of heart disease by about 20 percent or increases it by over 100 percent. Once more, this time with a logistic model, the result swings widely with sample size.

Finally, in figure 6.4 we have the relative risks and P-values as estimated by the hazard technique. These plots are a little different. In particular, we want to talk about the P-value or confusion index associated with each risk value and we will do that in just a moment after some other points are made. The top plot shows relative risks for drinkers and the bottom those for smokers. The overall patterns are about the same as on the prior figures, smoking is always bad but by very different amounts, and drinking may be good, bad, or indifferent. Also note that the spread of values is greater than what we had for the logistic version. Here, the relative risk for smokers runs from about 0.9:1 to 4:1. That's a change in risk from –10 to 400 percent! The sequencing information we introduced with those random time assignments does this. If we now selected another set of random numbers to assign a new time sequence to the subjects, a completely different picture would result. The peaks and valleys would change both in position and value. Clearly it is not simply how many people apparently die from smoking, but also the order in which they do it. Just how much difference the ordering can make is illustrated in figure 6.5, in which the same total of 129 subjects were analyzed for thirty different random orderings. It is probably not surprising to see the risk for drinking bouncing between good and bad, but notice that there is one ordering that even says

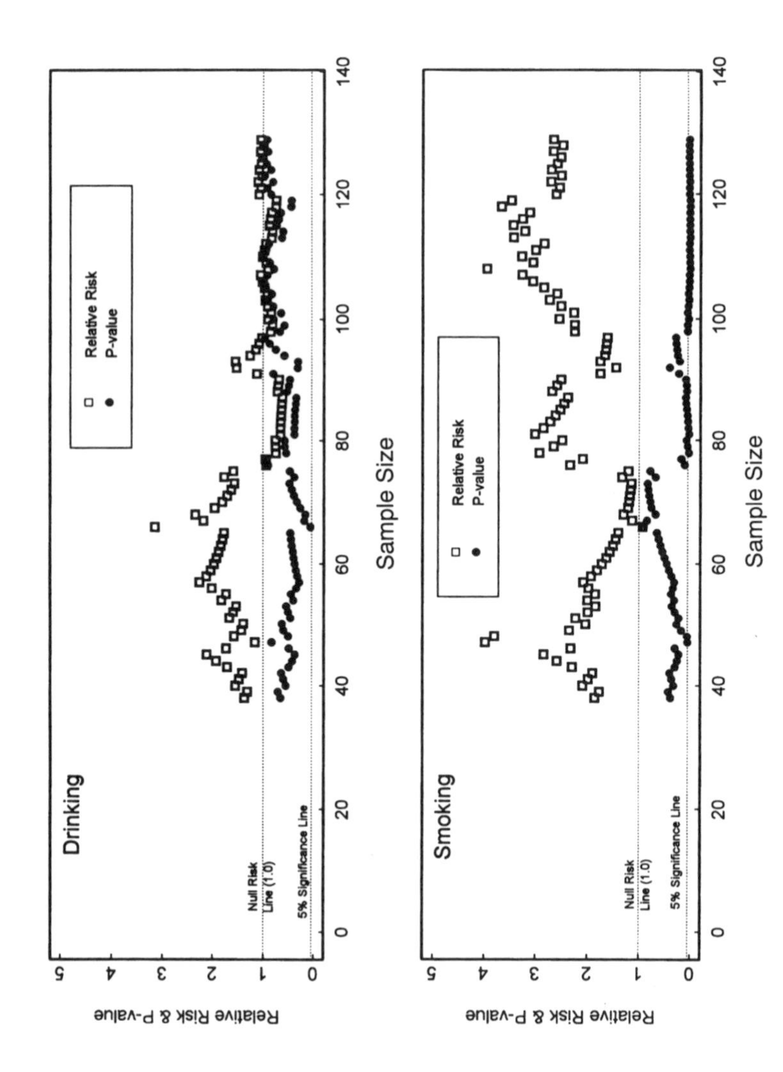

Figure 6.4: Drinking and Smoking, Hazard Relative Risks as Sample Size Increases

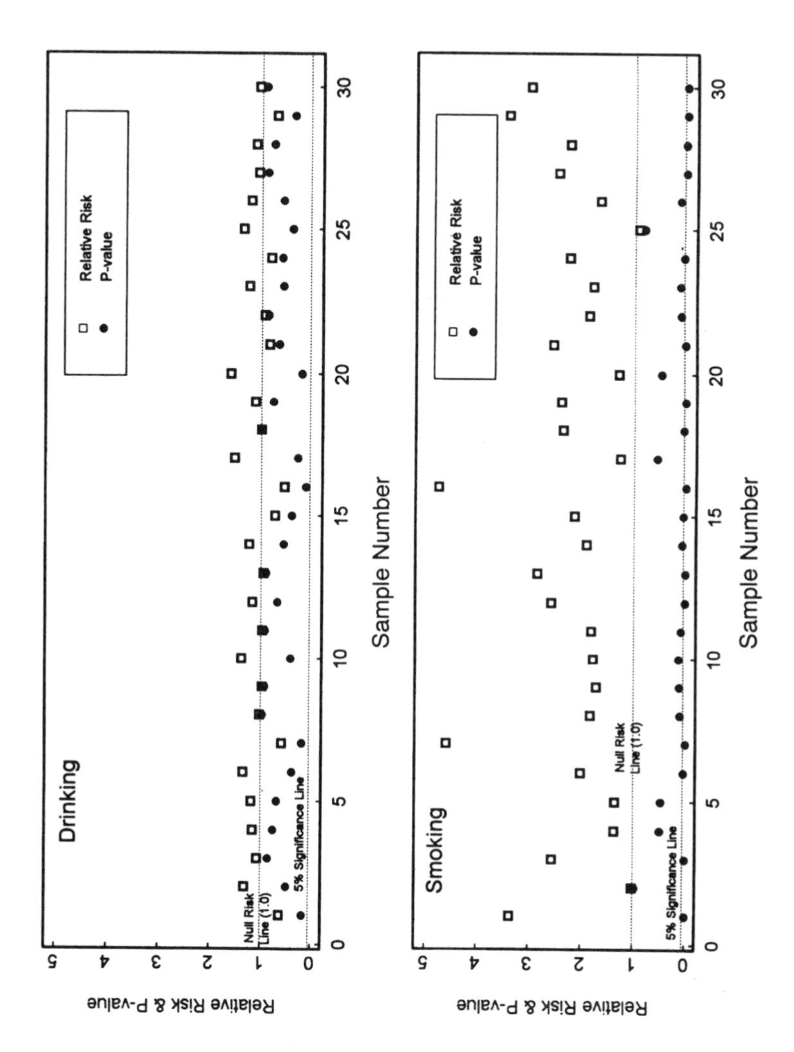

Figure 6.5: Drinking and Smoking, Hazard Relative Risks for 30 Random Orderings of Same Subjects

smoking is beneficial (the lower plot, at sample number twenty-five, where the square falls below the Null Risk Line), although, the P-value for that ordering is terrible. It is close to 1, far above the 5 percent significance line. That's a good example of the sample that would fall beyond predicted confidence bounds.

Many probably realize that when I changed the sample sizes for these plots, I must have obtained different confidence bounds and significance levels for all the different answers. That is correct. However, there is no need to jump to the conclusion that the statistical tests for acceptance will reject all the bad sample results. This is the reason for including the P-values in figures 6.4 and 6.5. For each of the estimated relative risks shown in the plots as open squares, the corresponding significance level is shown as a solid dot. Even at sample number sixteen in figure 6.5, the greatest relative risk for smokers, the significance level is 0.02 percent!

On both plots I drew a line at the 5 percent significance level so that a quick glance lets us pick out sample sizes that result in significance levels at least that good. Contrary to intuitive reasoning, a larger sample size does not guarantee a better confusion index (a lower P-value). The sequencing of events is all important in the hazard model.

One last note on the importance of sequence. For a really simple-minded case, assume that on successive days in an experiment, one test subject, A, died, then one control subject, B, died, then one test subject, and so on. The sequence of deaths is A, B, A, B, and so on. The relative (hazard) risk of test to controls turns out to be about 1.3:1, no matter how long the test goes on. Say it goes on for five years, with about 1,000 each of subjects A and B dying. Now, if for some reason the sequence was observed one day later (maybe somebody lost the record for day one), the sequence switches to B, A, B, A.The relative risk for test to control is now 0.76:1, the reciprocal of 1.3. It just depends on where we start the clock. And, as some have probably suspected, a logistic- or table-type model for both of these cases would say, correctly, that the relative risk is exactly 1.0; no difference between the test and controls. The hazard analysis would be wrong no matter which of the two results we happened to get. In lesson 7 we examine a real-life, high-profile study that completely collapsed because no one worried about when the clock started.

Even though the sequence data generating figure 6.4 is artificial (I made it up) it nonetheless sheds more light on the on-again-off-again pronouncements concerning risk factors. Now consider the drinkers described in that upper plot. The risk factors bounce up and down just as before, but look at the P-values. Only a few of the points are even near the 5 percent line, let alone below it. The message is loud and clear. There is no relation between drinking and heart disease. Any point that indicates otherwise is simply not significant (statistically). Unfortunately, researchers never examine their data in this manner so, if, when they end their test, the P-value happens to be one of those acceptable numbers, they truly believe that single result and clamor to publish it.

The results from three analysis methods and various sample sizes are summarized in table 6.5 where the maximum, minimum, and final values for all the combinations are presented. The final, or maximum sample size results are called the "study" results because normally these are the only ones we see or hear about. They are the only ones the statisticians let us compute. Remember sample integrity?

	Minimum	**Maximum**	**"Study" Value**
	Table RR (Odds)	Table RR (Odds)	Table RR (Odds)
Drinkers	0.46	1.86	0.96
Smokers	4.77	14.00	4.77
	Logistic RR (Odds)	Logistic RR (Odds)	Logistic RR (Odds)
Drinkers	0.75	3.78	0.96
Smokers	2.92	6.66	4.78
	Logistic RR	Logistic RR	Logistic RR
Drinkers	0.83	2.34	0.97
Smokers	1.86	3.03	2.08
	Hazard RR	Hazard RR	Hazard RR
Drinkers	0.62	3.15	1.04
Smokers	0.93	3.98	2.65

Table 6.5: Odds and Risks for Heart Disease Due to Drinking or Smoking (Estimated Relative Risks Calculated Four Ways and Using Different Sample Sizes)

So, entering the table at random, we could conclude that the odds ratio for drinkers to develop heart disease is 0.46:1 or 3.78:1, and that for smokers it is 2.9:1 or 14:1. Which do you prefer?

There is another very important point about these regression studies. I have said many times that it is imperative to include all pertinent variables. Once more I claim that it is *impossible* to know that all relevant variables have been included, simply because we cannot be sure they are all known. Many researchers seem to be aware of this. In looking over an assortment of seventeen studies on heart disease recently, I counted twenty-eight variables considered by one or more of them. They included things such as race, ethnic background, religion, coffee, education, aggressive personality (level), emotional stability, and fiber intake.* Some appeared to have negative effects, some positive. Until such time as all variables that ever exhibited an effect in anyone's study are included in the same study, none of these could be considered reliable. I wonder if the fellow who claims to have found over two hundred heart disease risk factors through the Framingham study mentioned in lesson 1 has run them all in one model. Until he does, none of his results is valid.

In general, it is difficult for an outsider to obtain copies of original data, but there are some available in textbooks and even on the Internet. Sheer physical volume, as well as proprietary considerations, mean that most available data are from fairly small studies. The largest I found had 195 cases, most only a few dozen. That is no real handicap but it might be interesting to explore the data from some of the grand studies involving hundreds of thousands of people especially if there were many cases in the group. In the following we consider some of the smaller ones.

The next-real life example has to do with a study of mortality of rats due to vaginal cancer.[3] Forty rats were divided into two groups of nineteen and twenty-one. Each group was treated with a different experimental preventive procedure, and then subjected to a powerful carcinogen. The time to death from the induced vaginal cancer was recorded for each rat. Because four of the rats were censored (for unknown reasons), the hazard method of data analysis is required. The analysis, following the death of the last noncensored rat, indicates that the relative risk of death from the carcinogen under one type of treatment is only 0.55, about 50 percent of the risk under the other type. The significance level or P-value associated with the result is 0.087. One suspects a slight beneficial effect due to the better treatment, but with a confusion index of 0.087, the result is not very convincing.

If we play the same sensitivity analysis games with these data as we did above, it develops that if there had been only nine rats, or if the

*Fortunately, in my opinion, few of these ever aroused the media, nor did I find any follow-up studies for many.

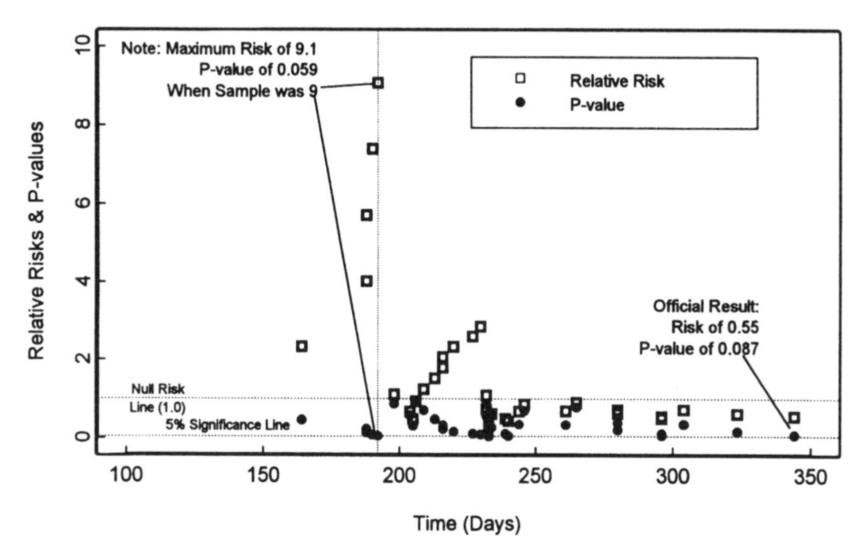

Figure 6.6: Carcinogenesis Data: Results of Analyses Conducted at Every Event after the Fourth (on Day 164)

---

study had terminated for some reason after the ninth event, the result would have been very different. After the ninth event the results show that that same treatment, rather than being beneficial, is now nine times (900 percent) worse than the other. Furthermore, the P-value is 0.059, still not great but "better" than the results from the full sample.

Figure 6.6 shows the results of an analysis at every event following the fourth (in the full sample of forty rats). The estimated relative risks are shown as open squares, the P-values as solid dots. The null risk and 5 percent significance levels are drawn in as before. Not surprisingly, stopping the experiment or limiting the supply of rats at an arbitrary number from five to forty would allow almost any conclusion desired, many of them supported by an acceptable significance level.

Let's pause a moment to see what we might conclude from these last several pages. Recall that the vaginal cancer data were taken from a real study, but the drinking-smoking data were made up to illustrate analysis problems. Though artificial, they do give rise to answers consistent with what we have witnessed on this subject in recent years. Using those artificial data to simulate a hypothetical first study on the subject demonstrates what vital information was lost by not considering changes in sample size. It is likely that expenditures of considerable time and money might have been avoided, as well as a lot of consumer confusion and more distrust of science. We know that many studies have been undertaken trying to pin down a risk for the drinkers. We also know that the controversy is still raging. It seems that each new test, in essence, winds up at a different little white square in the upper plot of figure 6.5. It's as though, in the long run, we will have a full study corresponding to each of those data points that were essentially lost by not being generated in the first test. We hope that (finally) at that point it will be decided that drinking is of no concern in heart disease. As for the effect of the experimental cancer treatment on the rats, it must be obvious that the experiment demonstrated absolutely nothing. Fortuitously, in this case, the P-value actually recorded, the last one, apparently was poor enough to discourage the researcher. There was no evidence that further tests of the particular treatment were undertaken. Even here, where there were evidently no substantial penalties, the researcher would no doubt have been even more certain of his conclusion had he examined sample size effects.

These exercises should give you a good idea of the problems underlying that tired old phrase: "More studies will be needed." You should have a feel now for what is actually going on, or what actually went wrong.

Many researchers know how to avoid the basic problems, such as too small a sample or a bad sample or use of the wrong model. More often, the traps are a little more subtle. One is that the variables used in the study have nothing to do with the problem.

That is not as strange as it might first sound. Recall the partial list I showed you from the twenty-eight variables various researchers had looked at for relations to heart disease. Examining large numbers, in fact all possible factors, for effects on the problem being studied is a must. Leaving out one that relates is actually worse than including nonrelated ones. Nevertheless, once it is decided that a factor is of no consequence, it should be eliminated from further analyses. These extraneous items, though not effective on, or even correlated with the disease of interest, do influence the curve-fitting process. They make the data "noisy" as some analysts like to say, and cause unnecessary errors in the results. The risks calculated for such factors tend to bounce around the value of 1.0, occasionally departing far enough to look interesting, and may even show up with an acceptable P-value. We saw a lot of that with the drinking variable. Of course, when this happens, the researcher believes he or she is "on to something" and becomes eager to publish the findings.

Another strange thing can happen when irrelevant guesses are included in the analysis. one or more of them can interact with real risk factors for the disease, resulting in spurious risks and incorrect risk values. If we once again think of the data as that cloud of points for the statistician's pen, think what happens if we "tossed in another handful of points," i.e., we add an assortment of random numbers. They can easily fall in a manner to visibly distort the general shape of that cloud. The curve-fitting routine in the computer will accommodate the new shape. As an example, I modified the data for the treatment for vaginal cancer in rats by interchanging the times of death for three treated rats with that of three nontreated rats. This resulted in a slightly reduced risk factor (i.e., the treatment seemed even better) and a greatly reduced significance level. I then introduced an artificial variable (noise) as a set of random numbers. A particular kind of random numbers, taken from one of those chi-square tables we have mentioned previously, were used. These are random, but the smallest number is larger than zero, no negative numbers are allowed, and they are skewed. That is, larger values have fewer occurrences. Think of looking at the set of test scores that are above an average of say 70. There are a lot in the 70 to 80 range, fewer in the 80 to 90, and even fewer in the 90 to 100 range. Our chi-square numbers are still random, in the sense of lack of structure, but they do exhibit a one-sided bias because there are no negative values. More importantly, they certainly do not bear any relation whatsoever to the variables in our example, and it is fair to refer to them as noise.

Computing the relative risks and the P-values for the treatment and the noise separately and together, we get the results shown in table 6.6.

	Treatment Effects		Noise Effects	
	Rel. Risk	P-Value	Rel. Risk	P-Value
Treat Only	0.43	0.016	—	—
Noise Only	—	—	0.94	0.138
Combined	0.33	0.003	0.90	0.022

Table 6.6: Effects of a Random Variable (Noise)

Note that the P-value for the true variable, treatment, by itself is near 1 percent and its relative risk about a half (0.43). For the noise alone, P is about 14 percent—meaningless of course, as is the relative risk for the noise, approximately 1.0. When the two are combined, the apparent risk for the treatment is somewhat better (0.33) but the P-value is even more convincing, almost ten times better (smaller). And note, the noise seems to have a mild beneficial effect, a relative risk of 0.90 and an interesting P-value near 2 percent. I hasten to add that not every set of random numbers has this sort of impact; the point is, it can, and so can other meaningless factors. Sorting out the truth in these situations is very difficult and it is easy to see why numerous independent studies that arrive at the same result should be made before the public begins receiving warnings and recommendations. Even though this sort of problem exits, all possible factors must be considered and these questions resolved for each of them. It is far worse to leave out a real risk factor than to include meaningless ones. The latter is a case of the lesser of two evils.

As a matter of interest, I calculated the correlation coefficients for all the variables in this example. The results are shown in table 6.7.

	Time	Death	Treatment	Noise
Time	1	–0.17	0.35	0.23
Death		1	–0.15	–0.15
Treatment			1	–0.18
Noise				1

Table 6.7: Correlations for Factors in Table 6.6

All of these values are small in my opinion. The value 0.35 for treatment and time is of interest and appears to confirm the reduced risk for the treated rats, but I would not put great import on that. The two values of –0.15 for death with both treatment and noise are a pure coincidence. A scatter plot of those two, one against the other, shows a mild trend and in a real study should certainly be pursued for an explanation. (Here it was a coincidence with the particular set of random numbers that came up.) Hopefully such an investigation, using information not otherwise relevant to the study, would indicate that the noise should be removed from the analysis.

Some textbooks, some papers by the statisticians, and common sense continually warn investigators (study makers) to examine the data. They encourage making all manner of pictures to display it.

The plea is to try to understand the data before any analyses have even begun. If you plan to undertake studies in the future these warnings should be taken to heart. In doing this, never be afraid to be creative in examining your numbers. Ask "What if?" then do it and see. A case in point, the game used here of plotting analysis results as a function of sample size, can be very informative. If the data displayed in figures 6.5 or 6.6 were from your experiment, would you be eager to report them in a major journal or the *New York Times*?

Now, about those studies with thousands and thousands of people involved. The thing they really have going for them is impressive volume. A closer look reveals that in most cases, there is little else. Sample selection for these regression studies is important for the very same reasons it is important in any survey. What is required is a representative sample. A simple way of stating it for these studies is that the relative proportions of people with and without the diseases, or dying and not dying, must be the same as in the population at large. In principle, this can be achieved with a fairly small number of subjects (compared to tens of thousands). Of course, if we study rare diseases, we do need a large sample simply to find the very few who come down with the malady. However, most large studies are done on popular diseases such as heart attacks simply because it's easier to get the big bucks and volunteers needed to do them. Many people probably contribute to the heart drive, but how about the Phister Syndrome Foundation for people with blue hair? Consider the following examples.

For the drinking-smoking case discussed above, the sample size was 129. If we think about increasing the sample, only two kinds of things can happen. If the 129 subjects are not representative and we select more from the real population, then the computed risks will continue to fluctuate until we reach a sufficient sample size. That is, the patterns in figure 6.4 will eventually stabilize if, in fact, we are selecting from a consistent population and doing an appropriate analysis. If however, the sample of 129 is representative, then increasing the number of properly selected subjects will not change the estimated risk values. Usually however, the confusion index will decrease as the sample size increases. Plots such as those in figure 6.4 instantly reveal when stability occurs. In most cases, a sample of a few hundred subjects is all that is required, assuming again that it really is representative. Of course, one can never be certain that such is the case.

Two of the more famous large and long-term studies here in the United States are the Nurses Cohort Study, run by the Harvard Medical School, and the Framingham Study, being conducted in Framingham, Massachusetts, under federal funding. Both have been under way for over twenty years now. The former enlisted more than 120,000 nurses around the country, the second began with about 5,000 residents of Framingham. Both groups comprise cohort studies for a host of diseases and health problems. Now, here are two very large samples. When we read about results from them we must be impressed. However, it is fair to wonder if these huge, expensive studies are worth their cost. Let's take a peek.

If we are to believe that results from these studies can be used to dictate treatments for anybody, the very first obvious question is this: Are nurses or Framingham residents representative of the whole United States? Another related, and simpler, question follows: Do nurses and Framingham residents represent each other? In other words, should we expect consistent results from the two? As it turns out, in at least one case, we don't even have to consider those questions; the experts running the studies produced conflicting results mostly because they made very different studies which were reviewed and reported as conflicting without even highlighting the differences. The public was led to believe that, once again, two groups of medical folk conducted the same test and came up with different answers.

In lesson 3 I mentioned that two reports on estrogen-heart disease studies, with opposite results, appeared in one issue of the *New England Journal of Medicine* in 1985.[4] Those reports came from the two studies just described. The Nurses Cohort group, using a hazard-type analysis, found the relative risk for cardiovascular disease for estrogen users to be only 30 percent of the risk for those not using estrogen. The other, the Framingham Study, using a logistic model, found the risk for estrogen users to be 76 percent greater than for nonusers.

Recall that the journal's editors were apparently disturbed, and called upon their statistical consultant for an explanation. As mentioned, his conclusion was "the investigator's great cop-out: More research is needed." In discussing differences in the procedures and models used, his comment was that "it does not seem likely to me that differences in . . . mode of analysis could, in themselves, cause these studies to point so strongly in opposite directions."[5] Why a statistical consultant was unable to believe that different models can produce different results is really quite a mystery, but that is what he said. He also overlooked two major differences in the studies, either one of which could have been responsible for the conflict: the ages of the subjects and inclusion of a known high risk factor in one study but not the other.

Both groups had a large number of variables to be examined: smoking, drinking, blood pressure, age, body fat index, and so on. The Nurses Cohort included parental history of heart disease as a factor, the Framingham Study ignored this. Now what have we heard about family history and the risk of heart disease? After all we have seen here about the necessity of including all factors in any study, how could anyone not expect different results for these two cases based on this discrepancy alone?

In addition, we all know that age is important in the incidence of heart problems. The Framingham folk even published a value of relative risk for age of 1.8. In other words, older people have more heart trouble and we could reasonably expect risks for other conditions/diseases to change with age. Well, the nurses group was relatively young compared to the Framingham group. The nurses ranged from thirty to fifty-nine years old, average about fifty; the Framingham ages ran from fifty to eighty-three, average about sixty-three. Sixty-two percent of the Framingham group were older than the oldest of the nurses. And still the reviewer was confused. Here is an example of a nonmathematical reason for the confusion we see. When the experts won't explain the most obvious of problems, what can we expect from the columnists?

By now, of course, most of us are well aware that different models can produce different results from the same data. It is interesting nevertheless to play games with this example. We took the Framingham data counts of heart trouble incidents (cardiovascular disease or CHD in their terms) in the two groups, users and nonusers of estrogen, and assigned random times to the events. This enabled use of a hazard model approach to their data. Recall that the Framingham data did not consider event sequencing; the study used a logistic model. This simple ploy accidentally produced a bad result for estrogen users, a relative risk of 1.21 versus the Framingham claim of 1.76. Then we arbitrarily changed the event times for the users by 46 percent, i.e., increased their survival time by that amount. The risk dropped to 0.87, more like the Nurses Cohort study. The resulting survival curves and relative risks, as given by a hazard model, are shown in figure 6.7. Our shift in the data shows up as the change in position of the solid line. Well, someone could say if we manipulate the data we can get any result we want. True, but the point is that in the absence of any event timing data for the Framingham Study to actually make a hazard analysis, it is fair to say that, were such data available, the result could have been anything. Comparing logistic model results to hazard model results is an apples and oranges game. There was no need for the great fuss that the *New England Journal* made of the issue. Without going back and doing the same type of analysis on both sets of data and throwing out factors that don't appear in both, we can never know if those studies are really in conflict.

Many more studies of the estrogen-heart disease relation have been completed since 1985. To this day there is still doubt in many minds as to whether hormone replacement therapy is advisable, but the consensus seems to be swinging in favor of the Nurses Cohort result. What is really amusing is that the promoters of estrogen willingly quote "exact" numbers for the relative risk. They say things like you are "37 percent better off by taking estrogen." It is truly amazing that so many practitioners take those less than exact numbers so seriously. In this case it would be far more truthful to say "there is about a one-third improvement from taking estrogen."

As we said, both these studies have a large number of subjects. However, for this particular result, the Harvard people were able to use only about 30,000 of the 120,000 nurses. The Framingham group had a mere 1,234 eligible female participants. It's obvious then that the Harvard result must be the right one, 30,000 versus 1,234? Well, these totals are nearly meaningless; what is really important is how many cases, deaths or heart attacks, actually occurred. The Harvard group had about 140 out of their 30,000; the Framingham folks about 200 out of their 1,234. (Keep in mind the older ages of the Framingham subjects.) By this measure of total cases, they are comparable.

The nurses study "needed" 30,000 subjects to achieve a suitable number of cases because the young ages, not the total count, per se, is important. To see this, we took that total number of 140 cases, held it fixed, and starting with a small group, added more and more noncase subjects to achieve larger samples. The smallest one was a total of only 360, with the relative risk for current users of estrogen to those who had never used it estimated at 0.52. Continually doubling the sample, up to a maximum of 5,120 subjects, moved the risk between 0.52 and 0.56. The confusion indices varied from 0.00006 to 0.00039! All of this confirms what we claimed above: For common diseases, thousands of subjects do no more than impress

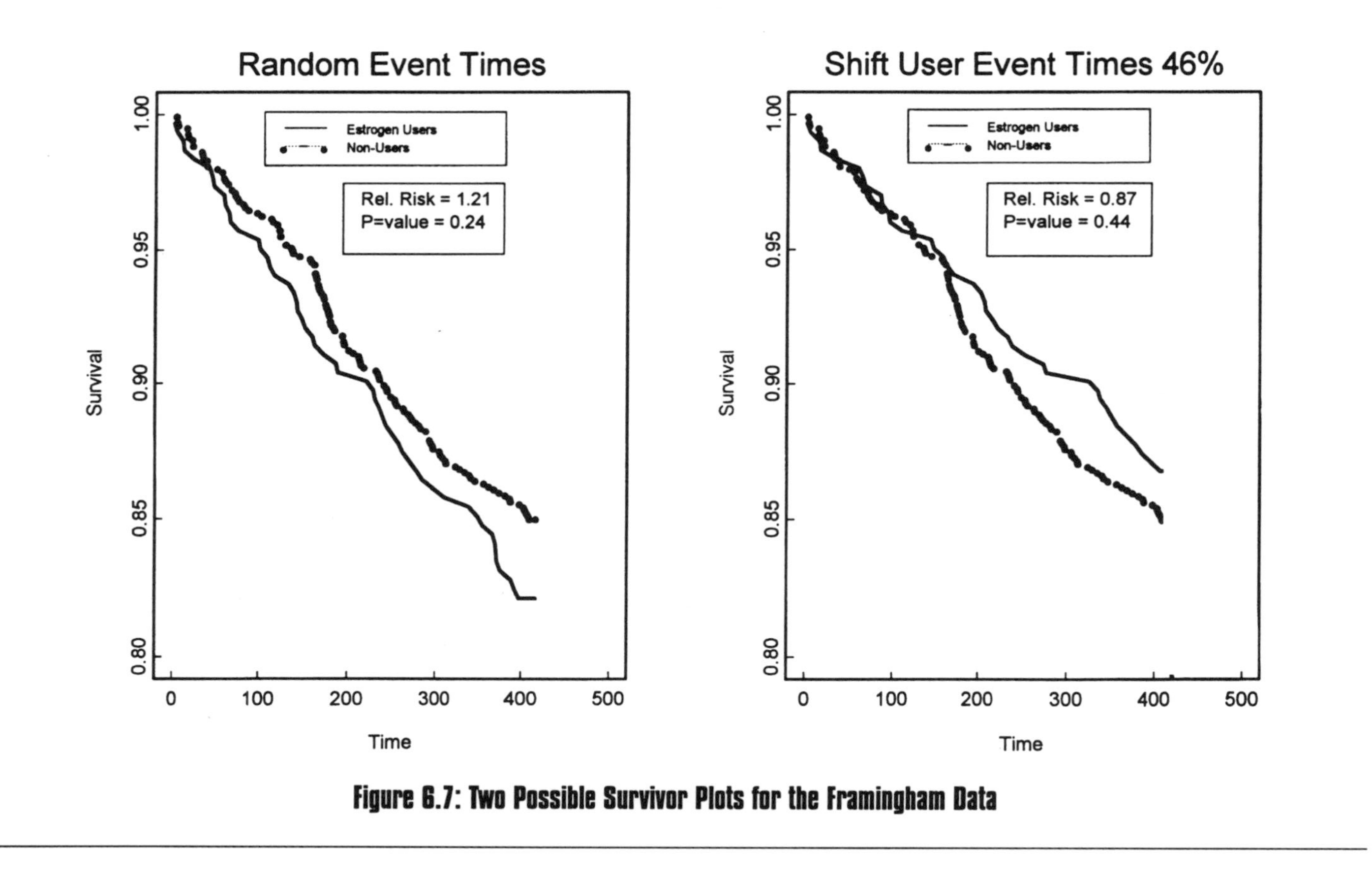

Figure 6.7: Two Possible Survivor Plots for the Framingham Data

the media and the public. A few hundred is all that is needed as long as the number of cases is reasonable. However, if we want to consider childhood leukemia and electromagnetic fields, that's another matter.

This lesson has illustrated many reasons for the conflicting study results. I trust that it will also make more people a little skeptical of the next report they hear or read about. The next lesson yields even more insight. We will see how the interested parties don't really comprehend what they are doing, and the experts dwell on theory, with little interest in usable results.

If you had no prior background in statistics, this lesson has introduced a lot of new concepts. We talked about three major classes of studies: table, logistic, and hazard, and tried to point out the significant differences in them. We showed how the table and logistic models produce similar kinds of answers with the advantage of the latter being that it can handle many variables without becoming unwieldy. We emphasized that the hazard approach is keyed to timing, or the sequence in which events occur, and gave an example of how it can be used even though all subjects die during the course of the study. Recall that the table or logistic models fail in such a case.

We also repeated, in many ways, that the integrity of any study is ultimately dependent on the inclusion of all possible factors. Even though firm indications and reliable trends may be possible without such complete knowledge, no quantitative answers can be believed. In recognizing this point, we must acknowledge that very few studies, beyond well controlled A-B comparisons of well defined conditions, can be taken as definitive. In most of the more interesting situations, for example, effects of diet, lifestyle, and so forth on heart disease or cancer, uncertainty will never reduce to zero. The controversies will continue to rage.

However, the overriding message of this lesson is the fact that the researchers frequently misuse the tools they so cherish—these models. It is not necessary to have the *New England Journal of Medicine* make an issue out of conflicting results to realize this, it is apparent in reading many reports. There is little appreciation for how difficult statistical studies really are and how misleading seemingly good results can be.

Unfortunately, we do not have access to the details of studies, we must settle for what the columnists report. We have however, suggested some guidelines for use when reading those reports. They can be summarized as a list of questions to ask:

- Is this one of many studies on this subject?
- Do the other studies agree with this?
- Were there at least several hundred cases (not just subjects) evaluated?
- In your judgment, were the subjects studied representative of you?
- If you know or can find out, did the other studies cover types of subjects different from this one?

If the answer to any of these questions is no, just tuck that article away for future reference, there is certainly no need change you habits or start eating the recommended diet.

Notes

1. D. A. S. Fraser, *Statistics: An Introduction* (New York: John Wiley & Sons, 1958), pp. 270–72; James J. Schlesselman and Paul D. Stolley, *Case-Control Studies: Design, Conduct, Analysis* (New York: Oxford University Press, 1982), pp. 227–87; and John D. Kalbfleisch and Ross L. Prentice, *The Statistical Analysis of Failure Time Data* (New York: John Wiley & Sons, 1980), pp. 70–142.
2. Klaus Hinkelmann and Oscar Kempthorne, *Design and Analysis of Experiments* (New York: John Wiley & Sons, 1994).
3. M. C. Pike, "A Method of Analysis of Certain Class of Experiments in Carcinogenesis," *Biometrics,* no. 22 (1966): 142–61.
4. Peter W. F. Wilson, Robert J. Garrison, and William P. Castelli, "Postmenopausal Estrogen Use, Cigarette Smoking, and Cardiovascular Morbidity in Women over 50," *New England Journal of Medicine* 313, no. 17 (October 1985): 1038–43; and Meir J. Stampfer et al., "Prospective Study of Postmenopausal Estrogen Therapy and Coronary Heart Disease," *New England Journal of Medicine* 313, no. 17 (October 1985): 1044–49.
5. John C. Bailer III, "When Research Results Are in Conflict," *New England Journal of Medicine* 313, no. 17 (October 1985): 1080–81.

Lesson 7

Do New Hearts Help? (A Review of a Defective Study)

Heart transplant surgery has been around for well over thirty years, and it is still something to marvel at. Just imagine the excitement in the late 1960s when the technique was brand new and every operation made headlines. Some of the byproducts of that excitement comprise the material for this lesson.

This fascinating subject provides us with many examples of the things that go wrong with studies and their reports. In this instance they range from an initial lack of planning and a forced response to a hot topic to a paucity of understanding of experimental design to the eagerness of theorists to grab headlines and ignore their own results. This case, which we will discuss in detail, dragged on unnecessarily for years, probably because of the glamour and potential impact of the subject. What should have been the final word was published, by competent people, within two years after basic problems were recognized.[1] The report, which appeared at about the time the official study ended, properly concluded that the longevity question could be answered either favorably or unfavorably, depending upon the choice of analysis, and so the data were inconclusive. The whole issue of transplanted hearts and whether they extend life should then have lain quietly in its grave. In spite of that report and the (then obvious) fact that the study was flawed and the data worthless, diehards continued to publish their "new" but equally useless views for nearly a decade. In a way, this was fortunate, because the study, together with all the subsequent research activity, now provides us with numerous vivid examples of why reports conflict.

Not surprisingly, transplantation did not meet with a lot of success in the early years. Even when the operation itself was successful, patients were not surviving all that long. They tended to die rather soon after surgery. This caused the transplant center at Baylor University, one of the two major ones in the United States (the other is located Stanford University) to close down in just a few years. At Baylor, it was not clear that the short extension of life was worth the extraordinary costs. The question remained open at Stanford.

What was needed, of course, was a study. A study would certainly demonstrate the benefits of transplantation. So in 1971, in spite of the fact that no real prepatory planning had been done, a group of the doctors at Stanford published an analysis of data from the records of the first thirty-four subjects.[2] This looked like a study, and the summary indicated real success: "In a select group . . . cardiac transplantation appears to prolong life." All the controls, those who did not receive a new heart, had died within three months, but a third of the transplant patients were still alive after twelve months. There could be no question, transplantation increased longevity.

The authors of that first report were obviously not well trained in the subtleties of statistical methods and this weakness came back to haunt them. Within a year, another doctor (who also happened to be a skilled statistician) published what proved to be a crushing critique of the first report.[3] His major objections pointed to fundamental problems in the entire experiment:

- Selection for transplantation was arbitrary, not random.
- There was no real control group.
- The defined apparent control group was badly biased. In fact, it was "demonstrated" to be essentially self-destructive.

A simplified argument for the last two objections is that the only reason any supposed control group existed was that those people were unlucky enough to die before a donor organ was found. As we shall see, however, it appears that some of the critic's reasoning was just as unsound as the study being criticized. Nevertheless, the researcher raised a real flap, and one wonders how, in a study of this importance, such fundamental tenets of statistics could have been ignored.

Well, as is often the case in the real world, when the Stanford transplant facility was established, no one seriously thought about making a long-term statistical study of the work to be done. If they had, much of the material described here never would have been available. As indicated above, it became apparent that there was urgent need for a study of transplant results when it was realized that patients were dying off faster than anyone liked. Often in these "emergency" studies, the only recourse is to scramble around, collect all the bits and pieces of information and try to examine them in some intelligent manner. Unhappily, we cannot impose critical statistical methods retroactively. If the data or their collection methods don't fit a recipe, almost nothing can be learned. Of course, the doctors who wrote the initial report were not statisticians, they were just using some statistical tools. They were not aware of all the shortcomings and subtle pitfalls, they were simply meeting the urgent need for documented results.

Normally, when a study is done, the questions to be examined are asked up front so that sound and relevant information can be sought. Defining the questions can be the most difficult part of any study. In these after-the-fact cases, the only recourse is to examine the data to see what questions might be answerable. (That, of course, is subjective hindsight, the grossest violation of statistical procedures.) If these questions don't satisfy the current need, and if that need is great, all kinds of wheels can be set in motion. Many people are inspired to jump into the act. That is what happened in this case. By 1977, the original transplant patient data had given rise to more than ten publications by over twenty-five authors. As a result, the bag of statistical tools was expanded and a raft of new researchers had their names in the journals in connection with very leading-edge subject matter. But the real questions were never answered. Let's take a look at what went on; it's instructive, at times amusing, and it might save all of us some future heartache.

As the critic pointed out, it was apparent from the first report that there was no planning for the data-collection process. The sample was not selected in any (statistically) rational manner and there was room for all kinds of bias. No amount of fancy arithmetic can correct for these problems. However, we can learn a lot by examining failures, so let's proceed.

The dataset is known as the Stanford Heart Transplant Data. It records events of the Stanford project from its inception in mid-1967 through early April 1974, is available in at least two publications,[4] and, currently anyway, can be obtained on-line from a producer of scientific software.[5] In addition, subsets (portions) of the full dataset appear in several papers. It is interesting to note that the actual numbers appearing in the various publications have several internal inconsistencies (other than simple typos) and they do not agree with each other. This is true even though one of the authors wrote papers presenting both the full dataset and one of the subsets. These conflicting "copies" is just one indication of the scurrying that must have gone on to create that first report. However, with some effort, many of the errors can be corrected or adjusted by forcing a degree of consistency. After doing that and making other changes that appear to be warranted after comparing the various versions, analytical results can be compared with published results and a close agreement to the averages of the published values can be obtained. That is, my results agree within a few percent of the published ones. I must, therefore, assume that my adjusted numbers are "correct." In any case, the remaining discrepancies have no effect on the spirit of this discussion. Unfortunately, the same cannot be said for discrepancies between the data subsets and the full datasets. If some of the subsets which were claimed to have been analyzed are extracted from the full ones and analyzed in the same manner, the results often disagree with those published. This is another hint of how much scrambling around actually took place. It would appear that many versions of the "data" were generated. I do not for a moment believe that to have been intentional, but it's certainly indicative of hasty or poor procedure.

Nevertheless, the data describe characteristics and events over the six and one-half year period relating to 103 candidates selected by a team of specialists for heart transplants. Potential subjects became candidates only if their condition was deemed hopeless by specific criteria and if certain logistic and consentual requirements were met. (Note that people are not selected randomly for heart transplant surgery.) The study interval for each of the patients begins with admission to the hospital and ends with death, departure from the study for various reasons, or when the patient is censored at the study completion date. Once admitted to the hospital, it is not completely clear what happens to the patients while awaiting a compatible donor organ. There was, of course, no way to know when such an organ might arrive and the reported waiting times extend to over a year. These waiting times become a very controversial issue.

At least nine kinds of data were collected, including whether a subject actually received a transplant, pertinent dates, and a number of medical evaluators having to do with organ transplants. In some cases, the subjects had undergone other cardiovascular surgery of some type. It was thought that perhaps that fact might influence the success of a transplant and so a record of prior surgery was included and later used as an analysis variable.

For patients who received a transplant, two distinct time segments are available. First is the waiting time for a suitable organ and second is the time to an event, either death or completion of the study. Each patient who received a transplant can appear in the dataset twice: once for the waiting interval and once for the time-to-event interval. As we shall see, the waiting interval is very nebulous. In fact, there is good evidence that all this factor did was obfuscate the already meager results.

In the first report, survival time for transplanted subjects was measured from the day of transplant; the waiting time was simply ignored. That was one of the procedural details focused upon by the critic. In his pseudoanalysis, he included the waiting time in the survival time of transplants, erroneously implying that the authors of the first paper had also included this time as a factor, and then he proceeded to criticize them for its inclusion. Perhaps he was trying too hard. Nevertheless, all successive reports distinguished the waiting intervals and treated them in still another way, as will be described below. It would seem that neither approach is justifiable in any sense of that word. Waiting times, as well as nontransplant survivor times, are meaningless quantities in this study. Why start a clock on the completely arbitrary day of admission to the study when many of these people had been critically ill (with heart trouble) for years? Why not start it, for example, on the day they first visited a cardiologist or the family doctor? At least those times have a similar quality. This is the first obvious problem with the entire dataset, yet it is never mentioned in the reports. The original authors ignored the waiting times, perhaps because there was no way to determine an accurate timeframe. All researchers who followed were so involved in responding to the critic that they never questioned the validity of any numbers. Recall that I said they all published different versions of *the same data,* never checking for consistency.

With the preceding material as background, this is how the story developed:

That first study paper of 1971 appeared as one in a series of rather matter-of-fact reports relating to the Stanford transplant project. At this early stage in the development of transplant research the series had an almost science fiction-like header, "Cardiac Transplantation in Man," and was devoted primarily to procedural and clinical details of interest only to the specialists. This particular report, although written by doctors, for doctors, had the word "prognosis" in the title and so was bound to attract wider public attention, particularly because this type of surgery was breaking new ground in medical science.

Most of the report was devoted to medical facts relating to individual subjects and the methods of selecting transplant candidates. The presentation of results is almost incidental: One graph had two survival curves showing the relationship of time to death for those who received a new heart and those who did not. This is the source of the survival numbers mentioned earlier.

In presenting the chart, the poor doctors, unwittingly it seems, did three things that the critics instantly seized upon. First, by showing the survival times for the two groups, transplants and nontransplants, the latter was automatically defined as the control group. The authors admitted that it was not a good one but argued weakly for its support. Then, in calculating survival time for the transplanted patients, the waiting period was ignored; only the time of survival after the operation was considered. Finally, two patients who showed great improvement while still in the waiting interval were released from the study and excluded from any survival calculations. Now these three things may seem rather innocuous, but problems come to light when we begin to think seriously about the questions addressed by the chart.

The graphic has two lines showing at what points in time people in the two groups died and the fraction of each group left after each death. It is most obvious that those with a transplant lived a lot longer than those without. Does that mean that transplantation makes for a longer lifespan? That is what the writers claimed. Well, that first critic seized upon the lack of statistical rigor in determining who was given a new heart. Besides the fact that the choices were (intentionally) not random, he argued that the only reason a patient did not receive a transplant was that the person died before a heart became available. Obviously then, it was the weaker patients who did not get transplanted and so the whole scheme was inadvertently biased to give new hearts only to those who were stronger, and they just naturally lived longer (with or without a new heart). In the critic's view, this point alone rendered the entire study worthless. Statistically speaking, it was a nonexperiment. He did not have to mention, as others later did, the problem raised by ignoring the two subjects who improved so much while waiting that they were discharged. By blatantly ignoring survival times of two people who were not transplanted, additional bias was produced in the original report. Ironically, if those two had not been ignored, it would have taken some of the steam out of the first critic's attack, because here were two nontransplant subjects who were still living—obviously they were not "weaker."

This criticism seems to have had the effect of spurring the interest of a number of statisticians. They wanted to see what could be extracted from this set of nonexperimental data from a real-life problem. (And not just any real-life problem, but one that had tremendous public impact.) An excellent report, published in 1974, included the waiting intervals of the transplanted subjects as well as the total times for the (by then) three nontransplanted subjects who had been ignored because they improved.[6] These authors argued that the waiting intervals should count as survival time of nontransplanted patients. The intervals were therefore included in total time allocated to nontransplants. Just about everyone thereafter followed suit when doing their own analyses. No one bothered to consider which approach had more or less bias or was more justifiable. They simply accepted and addressed, in this same manner, the first critic's comments.

Well, with these modifications it appeared that the survival times of transplanted patients were in fact no different than those of the nontransplanted. In statistical parlance, it was demonstrated that it could not be shown that the two were inconsistent. We must give credit where credit is due. The authors of the first widely published paper with that conclusion really worked hard. They applied six different models (none of which was a hazard type), that is, they analyzed the data in six different ways, candidly discussed the problems with each method and, as noted earlier, concluded correctly that with these data, the question of survival benefits from heart transplant must remain open.

Did that put an end to the matter? Of course not. Remember, this is high-tech medicine. Within the year another group, wondering if some factors might correlate with the longevity of transplants, got carried away with the arithmetic, essentially forgot what they were working on, and invented and published, as their title said, new methods for "tests of independence."[7] (But it wasn't really independence they talked about, it was correlation. Remember that in lesson 3 I said the terms are often mistakenly interchanged?)

In December of the same year another paper describing a new statistic invented just for the analysis of organ transplant data was published.[8] This author also used the drama of cardiac transplantation to display his new tool, which he had actually invented and published earlier in his Ph.D. thesis.

By 1977, the still fairly new hazard approach to modeling data with censored entries was brought into play by two authors who had written about this study earlier. Their work confirmed what had been said previously![9] They found that nothing conclusive could be said even though a few more variables were tossed in. This was deemed rather dull, so, eager to say something new, these authors came to the fascinating conclusion that the new hazard techniques appeared to be useful in transplant data analysis. And so it went. The data were even analyzed again, using the hazard approach but with a slightly different emphasis, as an example in a 1980 text on hazard modeling.[10] Even today, more than twenty years after the fact, the data can still be obtained on-line, as I have mentioned. The topic is still esoteric enough to attract the attention of researchers.

After all of the activity and varied analyses, by 1977, there was a consensus on the following points:

- The data simply did not contain any real information about the benefits of transplantation.
- The clinical procedures, i.e., patient care, seemed to improve over the six-year period.
- Patients' ages seemed to be important. Younger patients may have had a small advantage, but it was not clear how much.
- Patients with prior heart surgery may also have fared a little better than those without.
- Patients who waited longer for an available heart also seemed to fare slightly better than those who received one quickly. Longer waiting times meant longer survival.

The key question, "Does transplantation lead to a longer life?" remained unanswered. In addition, the liberal use of qualifiers ("seemed," "little better," "slightly," etc.) were also disappointing.

Even though we know that statistics can't prove anything, we still like to see words such as "strong indications" in study reports. The weak qualifiers show a lack of conviction on the part of the researchers.

So what really went wrong here? From the statistician's view, the major flaw was a total lack of planning. This was due partly to the fact that no one realized from the beginning that a study would be important and partly to the nature of clinical work. No researcher is free to manipulate all the variables in a manner fitting a statistical plan. How would anyone decide which transplant candidate to place in a control group (i.e., determine he would never get a transplant)? Then, how would that person be informed of his new status? In addition, the lack of statistical prowess among the authors of the first report allowed them to publish a paper that was fundamentally flawed. Granted, they may have succumbed to administrative pressure, but a skillful statistician could have done better, even if only in couching his results. This would appear to be another example of researchers believing their groundless claims simply because they don't know any better. It is not just the science reporters and the public who get confused.

It's easy to say there was no plan or that the right questions were not asked at the start, but what do those objections really mean? How do these omissions translate into specific problems? We already pointed to the basic problem here, the absence of a control group, and to the fact that there was no clearly delineated starting time to begin survival measurements. Are there still more problems? This is best answered by plowing through the data and identifying specific shortfalls that could have been avoided.

First though, let's dispose of our two initial problems. What about a control group? As the critic pointed out, the designated group may have been self-biased by being too weak to wait for a transplant. Several possible solutions could have been implemented if this problem had been considered from the beginning. In fact, the critic proposed three alternatives, one of which completely avoids any ethical preference questions (regarding assignment to a control group) in the random selection process. His method in this case involves pairing similar patients, randomly assigning a heart, when available, to one of the pair, and then measuring survival time for both, starting at the date of the operation.

The second problem, that survival time cannot be measured, is nasty and would have been avoided had there been a properly defined control group. As we stated above, for survival time after transplant, there is no problem, but when does the clock start for those not transplanted? All the analysts used the date of admission to the study as the starting time, but that was totally arbitrary. How long had the subjects suffered with bad hearts? How long had they been considered critical? The first report provides some information on these factors and it is seen that both those numbers varied from a few months to several years. Starting the clock at the hospital entry date is meaningless—one might just as well add random lengths of time to the stated survival periods.

A little later we will demonstrate that the "fix" that was used by the follow-up analysts merely confused the few issues that could have been more clearly addressed. No one stopped to question the first chosen solution. This is another common pitfall: someone stops to think, makes a statement, and then everyone else dons blinders and runs with the same ball.

So much for basic problems. Now, to better understand what went on in the various analyses, we naturally begin by looking at the data. The next several pages illustrate the kinds of things that should be done in data analysis but are invariably ignored—another reflection of poor training.

There are many ways to begin analyzing the data, and we will pick a few. We must acknowledge from the start that because these data do not come from a real experiment nothing quantitative can be said. All of the following is simply an exercise in how to examine data. Even in well-designed experiments we must hunt for surprises; the best of experimental designs will never cover all the bases. The surprises lead to the (inevitable) next experiment.

A good way to get a quick overview of this entire study is to plot the time history and end result for each subject. That sounds like a lot of pictures but can be condensed to one by using a scatter plot, as in figure 7.1. The date of admission for each subject is shown along the bottom, and the observation time for each subject is displayed vertically. Solid squares represent patients who died; open squares, the survivors, including the two early censorings. (The third early censored subject mentioned above died within a year after discharge and is recorded as one of the solid markers.) The slanted line is simply a time marker delineating all possible starting dates. Each point on it represents time zero for the subject plotted directly above it. The dotted line near the top is the closing date for data collection. All subjects on that line were censored at that point. The length of a line from the starting date to a square is the total

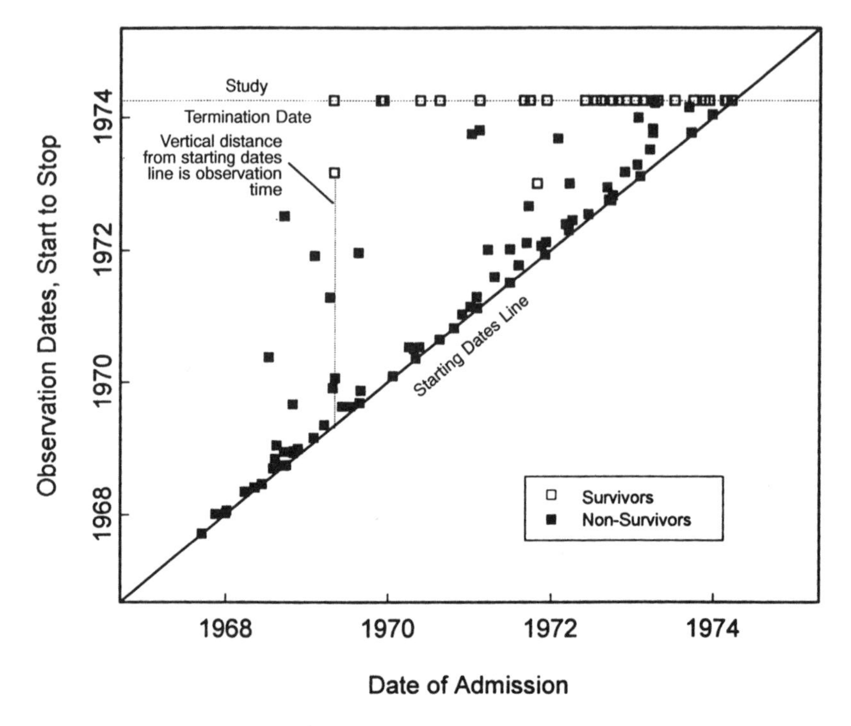

Figure 7.1: Subject Observation Times and Dispositions

observation time for that subject, including waiting times for the transplant cases.

The untrained but astute reader may look at the picture and think that something is obviously wrong. The people who entered the study later on, near the top right of the starting dates, weren't given enough time to die. Isn't that a kind of bias? After all, if we draw a parallel cutoff line, say at one year above the starting date line, a lot of those solid squares would become open ones at that line. Wouldn't that be more honest? Good question, and the beauty of a hazard-type analysis is that it avoids that apparent problem. Keeping the thought in mind, when we talk about survival curves a little later the explanation will be easy.

Now, with another picture similar to the one just shown, we can get a look at what all the waiting time squabble is about. Figure 7.2 is identical to 7.1 except that waiting times for the transplanted subjects have been delineated. They are marked with squares containing x's, generally clustered near the starting dates line. The visual

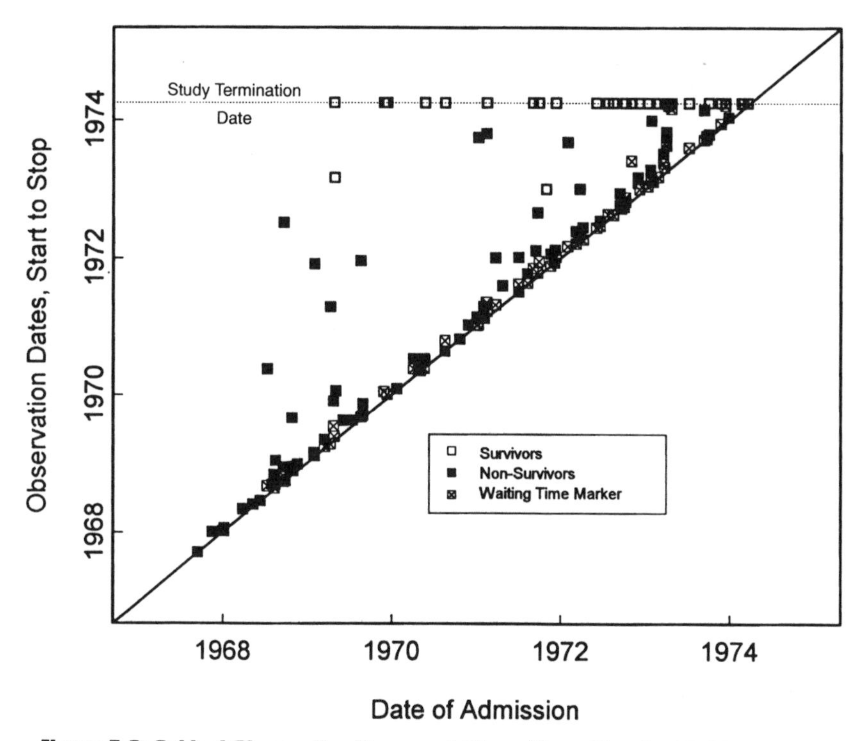

Figure 7.2: Subject Observation Times and Dispositions Showing Waiting Times

---

effect is not impressive. The added squares have little impact on the total picture. Notice however that they resemble the clustering of solid squares, mostly dead nontransplants, along the start line. Remember the critic's comment about the controls dying too soon? We should take a closer look. Are those waiting times similar to the times of early deaths?

An easy way to compare the waiting times with the early-death times is with a bar chart of waiting times and survival times. While we are at it, let's break out survival times for transplanted and non-transplanted patients. The result is depicted in figure 7.3.

Each bar in the chart has a width equivalent to 100 days. The height of the bar represents the count of all subjects with survival times within those 100 days. The total height of all columns is equal to the total number of subjects.

On the left we have the picture for transplanted subjects. That first tall bar indicates that about 45 percent of them survived for 100 days or less. That does not mean that they all died, many of those

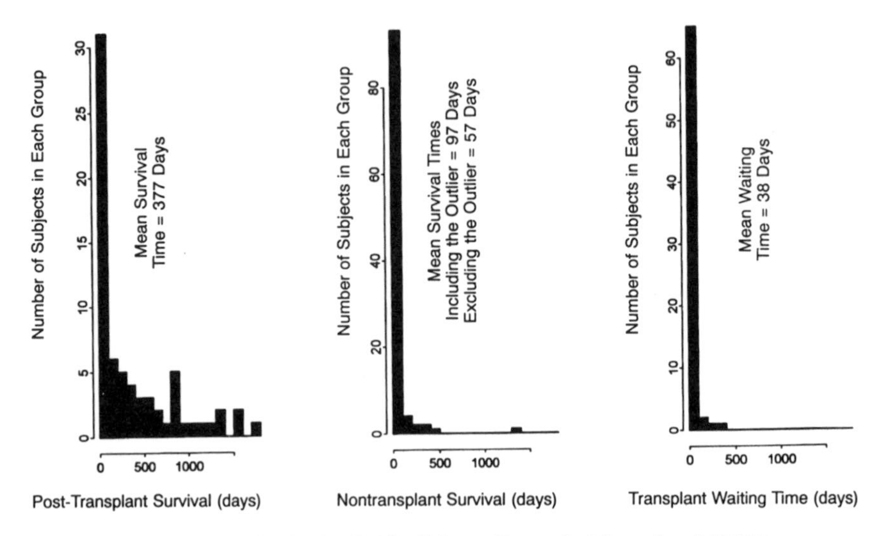

Figure 7.3: Bar Charts for Subject Times Group in Intervals of 100 Days

---

open squares at the top right of figure 7.1 are included in that bar. Other bars in this chart show fewer numbers of survivors for longer time periods. The very last bar indicates that one subject survived for a period of about 2,000 days or over five years. (The scale marker is deficient but that is the number for the last bar.) That subject would be the one represented by the upper left open square of figure 7.2. For comparison with the other plots, we made a note that the mean recorded survival time was 377 days.

The middle chart is for nontransplant subjects. Right away we note that it is not very dissimilar from the first. The blank area from 500 to near 2,000 might just be due to the fewer number of subjects. There were sixty-nine transplants in the first chart but only thirty-four nontransplants here, i.e., not enough to fill out the picture (but we don't know that, of course). That (possible) outlier on the far right of this plot is one of the two patients who improved so much that he was removed from the list. He was still alive at the end of the study with 1,401 days to his credit. That subject wound up generating considerable discussion. He is proof that some nontransplant subjects can survive for a long time, further weakening the case for transplantation. If we ignore him for a moment, the rest of the center bar chart then resembles the one at the right, waiting times. Even the average values of those second two pictures are fairly close. This supports the similarity of clustering we saw in figure 7.2: Waiting times are very similar to total survival times of (weaker?) nontransplanted subjects.

So, if we want to argue that the long-term survivor in the middle plot is really an outlier, for reasons unknown, the distributions of waiting times and nontransplant survival times are very similar, as the critic implied. But of course that could also indicate that the transplants were made "just in time" and the extended survival time is a true benefit of the operation. On the other hand, if we include the outlier in the center display, then it resembles the distribution of transplant survivor times, and transplantation is of no value. Clearly these preliminary plots are exposing a need for serious examination of the question. We may or may not be able to answer it.

Here we have an easily recognized example of two points: First, it demonstrates how just one outlier can raise havoc with study conclusions. Secondly, it is a vivid example of how the lack of planning came back to bite the researchers. There was no provision to handle this subject, no consideration had been given to what was really meant by "survivor time" and this very significant 1,400-day period remains as a red herring in the entire data base.

One of the things to do early on when exploring a dataset is to look for trends of any kind. They can illuminate inadvertent biases and uncover unsuspected relations. There are many possibilities here. For example, did the ages of selected subjects change with time; did age affect survival? Were all things constant, e.g., did techniques change, get better or worse, as the study progressed? Did overall success change with time? Perhaps we could think of other questions that might benefit from a quick look at a trend line.

As for age, figure 7.4 shows a number of trend lines relating subjects' ages to several variables as the study progressed. The three charts apply first to all subjects, then to transplanted subjects, and last to nontransplants. The top chart contains three lines, one showing the average ages of all subjects and one each for those who did and did not receive transplants. It is obvious that overall subject age was fairly constant, but as time progressed, the age of transplanted subjects moved downward. It appears that in the beginning, older patients were favored when doling out available hearts but preferences changed. In the center chart, we find that of those who received transplants, the older ones did not survive as well as the younger. The average age of survivors is always less than that of nonsurvivors. However, the age of survivors does not fall as rapidly as age for the group (that line is not angled as sharply as the one indi-

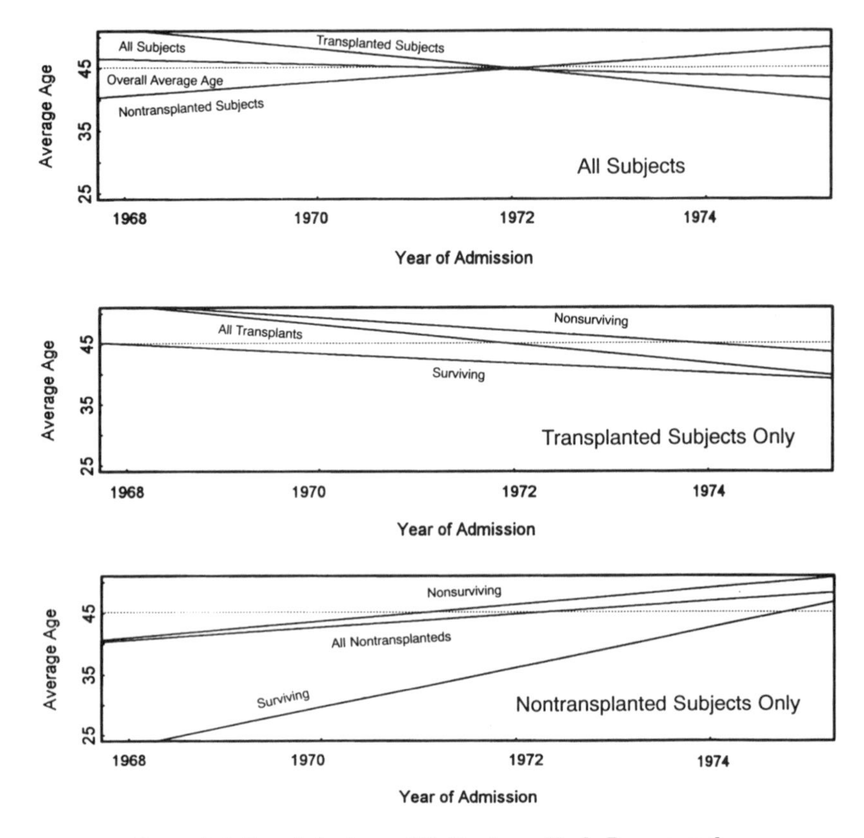

Figure 7.4: Trends in Ages of Subjects as Study Progressed

cating average age), which means that relative survivor age increased. That is, as the study progressed older subjects fared better. But we can't put any stock in these lines. They are simply straight-line approximations as to what really happened. The pictures serve only to guide and help understand the more formal modeling and regression procedures.

At the bottom of figure 7.4, we see that the average age of those nontransplant patients who died increased nearly in unison with all the subjects in the group, while the age of the survivors, those who were alive at the end of the study, jumped dramatically. However, the line for survivors is based on only four subjects and two of them did not enter the study until the last month. Their entry dates severely weight the right side of that line.

From figure 7.4 we find then, that as the study progressed

1. Average age was constant.
2. Older subjects were favored for transplants in the beginning, but younger ones were favored toward the end.
3. The age of survivors, transplanted or not, increased with time.
4. The age of nonsurvivors decreased for transplant subjects and increased for the others, but in step with the age of all subjects in those groups. It seems that, age-wise, a constant percentage of the nontransplants died.

Note that number 2 in this list appears to disagree with the conjecture that stronger patients preferentially received transplants, because older subjects should in general be the weaker ones. The team doing the transplants did insist that if there was any intentional favoritism in selecting heart recipients, it was for the weaker subjects. Perhaps there is more to this question of selection bias than meets the eye.

Number 3 supports the observation reported above that clinical procedures appeared to improve with time. It seems that everyone did better as time progressed. However, once more we must remind ourselves that these are qualitative observations only. No analysis has been done.

Next, let's look at those waiting times. They are of interest because of the way they affect apparent success of the transplantation process when they are either included or excluded from the survival computations. Before delving into any analyses, we should see whether the times are consistent during the course of the study. They are shown in figure 7.5, which has two scatter plots and corresponding approximate fits for survivors and nonsurvivors. We instantly observe that a few waiting intervals toward the end of the study were very long. The effect is seen for both survivors and nonsurvivors. Otherwise, things seem rather consistent throughout the entire six years. A closer look however, reveals that there is a tendency for the surviving subjects to have experienced a longer waiting interval than those who died. That is evident by noting that the dashed line falls above the solid one. That observation could also be taken to support the argument that preferences favored the weaker subjects, they did not have to wait as long, but died anyway. On the other hand it also lends indirect support to the criticism that the

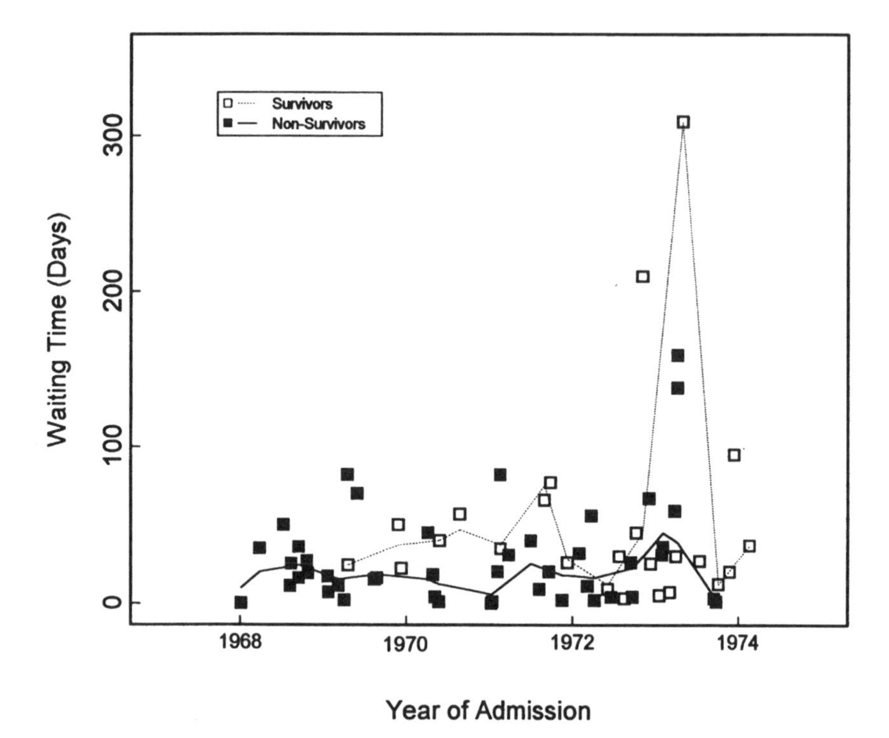

Figure 7.5: Waiting Times for Survivors and Nonsurvivors

---

weakest patients died before a heart became available. This figure, together with the center chart of figure 7.4, suggests an interesting positive relation between longer waiting time and greater survival. How that should be interpreted is not at all clear. Which precedes the other? Maybe neither comes first. Once again the ill-defined waiting time becomes the basis of more confusion.

Consider again the trend lines in figure 7.4. While indicative of some gross trends over the six years, they lack any touch of detail. A finer look is obtained if one considers taking a snapshot of the status of various groups at each event time. One kind of snapshot would depict the probability of death for members of certain groups. This can be estimated as the fraction of each group that was dead at various points in time. Such a calculation ignores any time or sequence information present in the data and looks only at the total number of deaths to that point.

As an example, we placed subjects thirty to fifty-five years old into four groups of five years each and distinguished between trans-

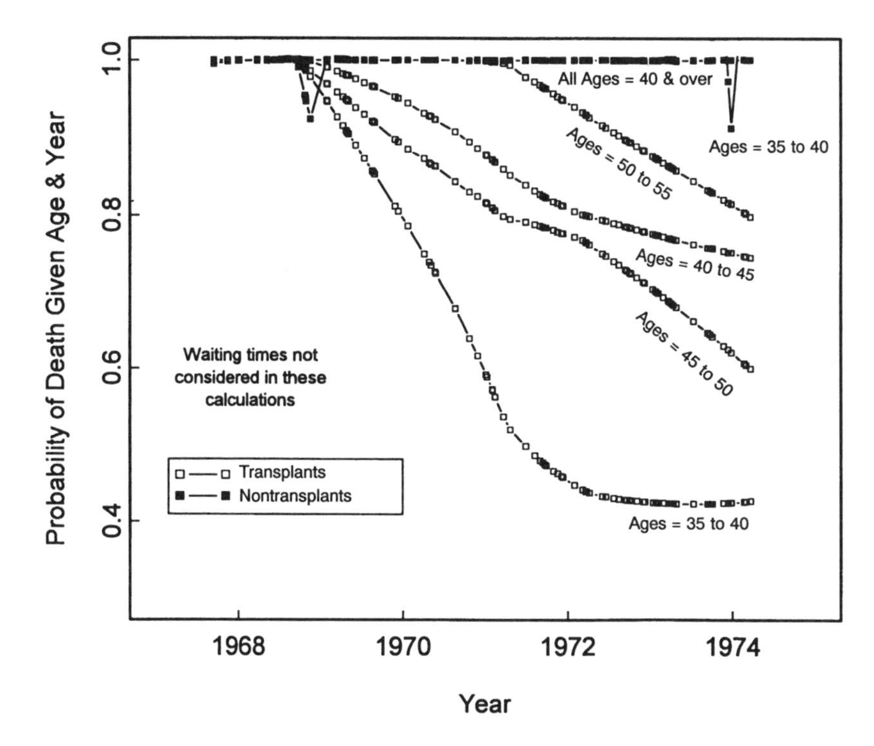

Figure 7.6: Probability of Death for Selected Groups and Years by Transplant Status

planted and nontransplanted. The estimated probability of a death in one of these groups is plotted, across six years (1968 to 1974), in figure 7.6.

This is the clearest picture so far of the advantage of youth and of procedural improvements with time. The chance of death is much less for the younger groups and it appears that nearly everyone's prospects were better later in the study. Notice that until sometime in 1969, everyone died, and that by 1972, only about 40 percent (probability of 0.4) of the young transplanted patients died. One might even venture a guess that a decisive change in care or procedures occurred at about the beginning of 1970 because all the dotted lines break away from the line for probability of one about there. It appears as though something significant happened at that time. A word of caution here. None of the sample sizes in the four age groups is very large. They range from six to twenty-one members, so the probability estimates must be taken with a grain of salt. The overall picture is probably okay, however, for seeing what is going on.

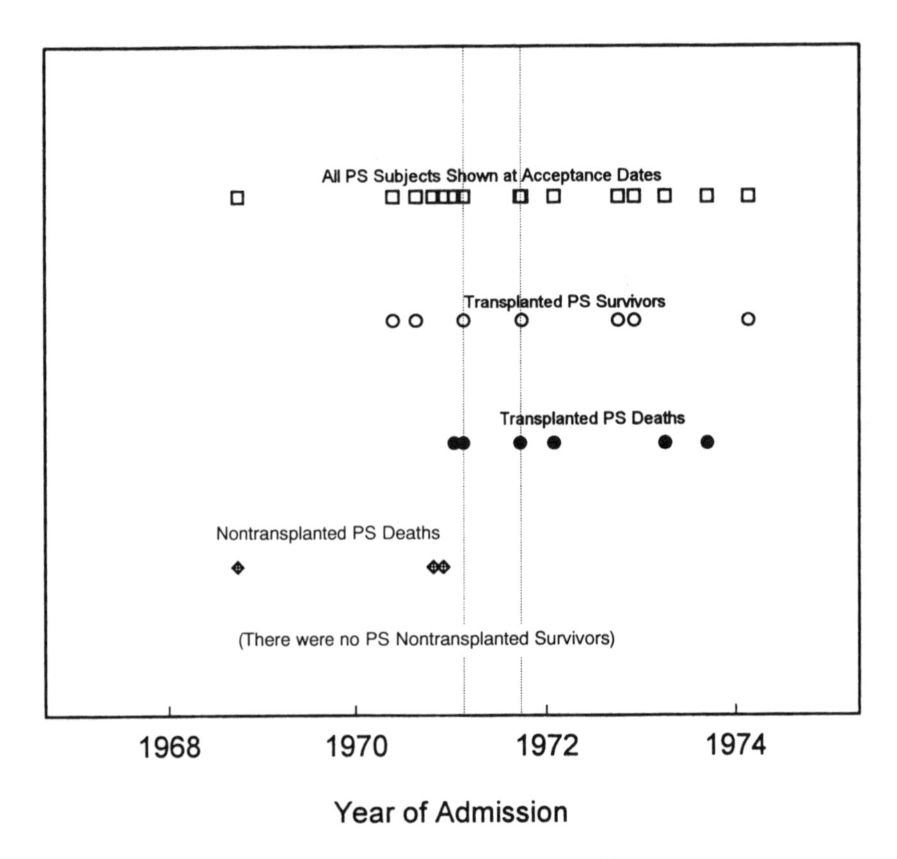

Figure 7.7: Subjects with Prior Surgery (PS): Summary

Because of the expressed interest in patients with prior surgery (PS) it wouldn't hurt to have a look at what happened to these subjects. Figure 7.7 summarizes their significant events. Each row of dots corresponds to a different type of event and the dots are placed at the acceptance date of the individual considered. There was a total of sixteen PS candidates, coincidence obscures two of them in the upper row. The overlaps occurred at the dashed vertical lines. The members of this (PS) group who had transplants, survived, etc., are available by inspection and counting. Eighty-one percent received transplants and 46 percent (six of thirteen) of those died. All three nontransplanted candidates failed to survive. Now, because less than one-half of the transplants died and all the nontransplants died, an analysis (detailed later) will show prior surgery as a favorable risk factor. But note that all the nontransplants' deaths occurred in the early part of the program when mortality was high for everyone, as we saw in the last figure. The analytical result is probably anomalous, but this picture is about the only way to discover that. Without it, one could easily publish the "fact" that prior surgery reduces the risk in transplantation. And recall that very few researchers like pictures.

There are a host of other plots we could examine to further explore the dataset. We could plot age against waiting time for various subgroups to look for selection biases. And of course, we could open the bag of all the medical evaluators that we have only mentioned in passing and explore them singly and in combination with the others. We are intentionally ignoring these indicators because we choose to concentrate on the factors considered most in the published papers. Naturally, any worthwhile evaluation of the data would include them.

By now we should have a feel for the kinds of areas to examine in a preliminary analysis such as this, and for some of the benefits derived, such as identifying unusual results. This is one way we get to know the data, something we cannot achieve by just running our number cruncher. We must also realize that many of the questions raised here could have been avoided if the experiment had been planned instead of being an after-the-fact look. The problems caused by the age trend lines, the wildly varying waiting times, and the biased placement of patients with prior surgery could have been avoided.

Having completed that preliminary review and feeling a little friendly toward the data, we can now fire up the high-powered software and pretend to get quantitative. Caution: Never fall in love with numbers.

First, let's review our preliminary observations. We suspect the following:

1. Age is important. Younger patients do better. (See figure 7.6.)
2. The clinical procedures improved over the six-year period so date of admission is important. (See figures 7.1 and 7.4.)
3. Patients with prior heart surgery seem no different than others. About half of the transplanted PS patients died and the few nontransplanted PS patients all died. (See figures 7.6 and 7.7.)
4. Longer waiting times make for better survival. (See figure 7.5.)

Some formal analyses should verify or refute these assessments and yield relative risk values where they may be important. But watch out for those expected anomalous results. Unhappily, this step of running the software is the point where most analysts start. They assume they know what to look for and that numbers are more believable than pictures.

For notational purposes we identify the following variables:

A = *Age* of the subject
S = Subject had undergone prior cardiovascular *surgery*
T = a *transplant* had been done
W = *Waiting* time
Y = *Year,* or date, of admission
DT = *Date* of *transplant*
AT = *Age* at *transplant*

A hazard-type analysis is undertaken in order to track sequences of events and allow for censoring. To make things a little more interesting, I will compare my results using the full dataset of 103 subjects, to similar results published in 1977, and to results for a dataset of only eighty-two subjects, compiled while the study was still in progress and published in 1974. (Recall that because of inconsistencies, it was not possible to replicate the results exactly.) The published results for the partial dataset were obtained using analyses other than a hazard type, but the essence of comparable results is the same. By the way, all the referenced analyses treat the waiting intervals as survival times for nontransplanted subjects in deference to the critic and the subsequent work. All the results are summarized in tables listing the study, the variables, and the computed relative risk for each.

First, the apparent overall worth of performing a transplant (T) can be shown in the following table:

Analysis	Variable	Relative Risk
Mine	T	1.08
1977	T	1.04
1974	T	0.88

Table 7.1: Relative Risk of Dying within the Study Period Transplant vs. Nontransplant

My result of 1.08, on the upper right, indicating that the risk of death with a transplant is 8 percent higher than without one, differs by about 4 percent from the published one below it. What should be noted is that according to the published risk for the full dataset, the operation apparently makes little or no difference in life expectancy. The relative risk is essentially 1.0, meaning no effect. However, the partial set, the 1974 study, claims that the risk after transplant is only 88 percent of what it was before, a 12 percent improvement. The authors of the 1977 report, who thought that the 1.04 value was good, noted the difference (one author was common to both papers) and suggested that it was probably due to the larger sample in 1977. The fact that changing the sample size reversed the outcome did not bother them. In lesson 6 we saw many examples of how results can fluctuate with sample size. Researchers are generally aware of this, and perhaps too much so. It is not often that they obtain results within one study, based on different sample sizes, as was the case here. When it happens and the results conflict, it is too easy to explain that with the remark, "it is only because there was a larger sample." The event should be a red flag and cause for deep concern until thoroughly understood. Ignoring it is a sign of poor procedure.

The next comparison is for the variable age (A) at time of admission, together with the transplant variable. This is not an interaction term of age and transplant acting in concert; including age here merely lets that variable account for some of the results, instead of putting all the blame (or benefit) on transplantation alone.

Analysis	Variable	Rel. Risk	Variable	Rel. Risk
Mine	T	0.99	A	1.03
1977	T	0.90	A	1.03
1974	T	0.71	A	1.03

Table 7.2: How Risk Changes with Age (Variables = Transplant and Age)

A look at that last column shows everybody agreeing on the effect of age, but now the transplant risk from the full group results Mine and 1977 has switched and become beneficial. If that seems strange, recall the relative chances of death for the various age groups in figure 7.6. The younger patients fared much better than the older ones. It appears in this analysis that age "absorbed" the risk of death, removing it from the transplant variable, so causing the observed switch in the two tables. The positive value of transplantation seen here also agrees with the impression from figure 7.6 when the solid and dashed lines are compared. For nontransplanted patients, the chance of death was almost always 1.0. Reversals such as the different results (shown in tables 7.1 and 7.2) when new variables are added to a hazard study are expected. The task of the analyst is to sort out the meaningful values. And, you guessed it, this is seldom possible and is one more reason that this month's report contradicts last month's.

We need to explain something about that age risk. The single number shown does not apply directly for everyone's age. Because the number is greater than 1 it does say that the risk increases with age but the risk for a specific age must be estimated by using the logarithm of that number. This is a mathematical detail for our purposes here. (Some researchers publish the logarithm of the number, others publish the number. It is a matter of taste.) What the mathematics do is move people up and down along a very steeply rising curve so the risk will get bigger for older people. We'll give some clearer numbers a little further along. All variables that cover a whole range of numbers, such as age or date, are treated in this manner. In any case, tossing age into the ring seems to have reversed the call from bad to good for the benefits of heart transplantation. Let's see what other variables might do.

How about the waiting time (W)? That seems as though it could be important, after all, lying around waiting for up to a year or so might not be healthy. It looks like this:

Analysis	Variable	Rel. Risk	Variable	Rel. Risk
Mine	T	1.10	W	1.00
1977	T	1.00	W	1.00
1974	T	1.10	W	0.99

Table 7.3: Is a Long Wait Bad?
(Variables = Transplant and Waiting Time)

And it seems to matter very little. All risks are now close to 1.0. (In my opinion, 1.10 or 10 percent is not really exciting.) This table says that neither the waiting time nor the operation matters. But this conflicts with our earlier impression, from figure 7.5, that longer waits made for greater survival. We begin to see how messy all of this becomes. Oh well, let's continue, we have two more variables to compare.

Next is the patient's actual age at the time of the operation (AT). This is not the same as age at admission because of those very long waiting times:

Analysis	Variable	Rel. Risk	Variable	Rel. Risk
Mine	T	0.09	AT	1.05
1977	T	0.07	AT	1.06
1974	T	0.09	AT	1.05

Table 7.4: A Small Change in the Age Used in the Calculation (Variables = Transplant and Age at Transplant)

Now then, look at that third column! The risk for transplantation is suddenly reduced by an order of magnitude, from around 0.90 in table 7.2 when computed with age at admission to 0.09 now. Does that mean that hanging around for a while and getting older, as opposed to just waiting, is somehow beneficial? In addition, even age itself now seems a little more significant than it was at time of acceptance. That risk moved from 1.03 in table 7.2 to 1.06 here. If all of this seems complicated and confusing, don't feel bad. Reflect a little on what is happening. In the first change we noted, from table 7.1 to table 7.2, when the benefits of transplantation appeared to reverse, the explanation was that this was consistent with the plots in figure 7.6. In table 7.1 we had considered nothing but the one variable, heart transplantation, either yes or no. In this study, many of the subjects died and the "blame" had to go somewhere. In effect, the model had no choice but to claim that the operation was not advisable. As soon as we added age to the pot, the model detected that this new variable was more to blame than the operation and reversed the outcome. Similar happenings occurred each time we introduced a new variable, or otherwise changed the mix. The model behaves as though there is a fixed amount of credits, good and bad, to be shared among the variables allowed. One can think of the trend lines of figure 7.4. Any time a variable shows a trend in a plot such as that, it would force the model to assign some credit to it. Only if the trend line were flat (no trend) would the model not be affected. This is the sort of thing that is happening here. It is the task of the researcher to try to include all the proper variables and so achieve a "correct" answer for each. As we have pointed out, variables may show trends, and so soak up credits in a model, even though they are not real risks. Our witch doctor fell into that trap and blamed daylight for hot weather.

Earlier I said that the mathematics associated with the age risk factor moved people along a very steeply rising curve and we would show this later. Now is a good time to do that. To use those numbers near 1.0 in the tables for age risk, we have to take their logarithms, then multiply by age and convert back again to get the true risk number. If we use the 1.03 value from table 7.2 and translate that into risks over the range of ages actually encountered in the study, we get the following:

Age	Relative Risk	Risk as a Percentage
8	1.3	130%
30	2.6	260%
45	4.1	410%
55	5.7	570%
64	7.5	750%

As we see, age can be really significant. It causes the risk to change about sevenfold as the patient gets older (according to this study).

The fact that those risk numbers rise so rapidly with age is an artifact of the hazard type of model and may or may not reflect reality. The fast rise is a consequence of the kind of equation, an exponential, that is used in the model. The choice of an exponential was not arbitrary, it was selected because of many other desirable characteristics. Even though such a rapid rise may be unrealistic, there are greater problems with alternative choices. The shortcoming is generally recognized by researchers but almost always ignored, just because there is "nothing we can do about that." That last statement means that we are "stuck" with this model.

As we have said before, all this modeling business boils down to trying to draw the best possible line through a scatter plot of measured events. In this case it is a plot of the chances of dying at each age. If we had complete freedom of choice, we might choose a straight-line (linear) equation to describe the way the risk increases with age. The results could be very different. To see how different, make believe that the numbers in the table above were the actual measured events but we choose a straight line instead of a curved one. Depending on just how we draw the line or, as we say, fit the curve, we could make it come close the first and last points or perhaps just to a central point. In the first case, for a fit at the youngest and oldest ages, where the risks are 130 and 750 percent, the forty-five-year-old, according to the straight line, would be assigned a risk of 550 instead of 410. Big deal? Well, suppose that the line were fit at the average age of forty-five, with the value of 410. Then, the oldest person would have a risk of only 610, and that eight-year-old's risk would be zero. So, depending on your choice of curve and how you fit it, you could pick estimated risks from zero to 750 percent to write about.

Again, a simple scatter plot would unveil the true nature of the change in risk with age and guide the choice of the kind of line to draw. The scatter would show us if the fast-rising exponential from the hazard model was even a fair approximation. If not, we could at least qualify the results. But, as I said earlier, few researchers believe in plots.

Finally, we can investigate the supposed clinical improvements over time as indicated in figure 7.6 by looking at the risks for the variable DT, date of transplant. In this case we have the following results:

Analysis	Variable	Rel. Risk	Variable	Rel. Risk
Mine	T	1.67	DT	0.88
1977	T	1.55	DT	1.00
1974	T	0.82	DT	0.75

Table 7.5: Did Things Get Better with Time? (Variables = Transplant & Date of Transplant)

And now, see the third column, the benefit of transplantation has reversed again, your chance of dying after a transplant is about 150 percent higher than if you don't have one. At least that's what the 1977 dataset says. And again, the smaller sample contradicts it. As for the date of the operation, the risk, in the last column, is a real mixed bag, it varies quite a bit. There is no reason for that other than different models and sample sizes, but those two things—model type and sample size—frequently explain why study results vary.

This detail should provide us with a very clear idea of why reports conflict. These kinds of changes in results are typical. Maybe now we can believe the statement, "More work will be necessary. . . ."

Notice that I haven't mentioned anything about the confusion index or P-values for any of these results. Do we really want more confusion? Seriously, for the numbers in the preceding tables, most of the associated values were less than 0.1 or 10 percent—even those that directly contradicted each other. Models aren't perfect, as we've learned.

The conflicting, confusing kinds of results shown here do not arise just because there wasn't any real control group in the study. It happens often in the multisomething studies. The textbooks now say that the statistician must rely on consultation and words of wisdom from the experts in the field under study, and on other nonmathematical, environmental facts and conditions. In other words, the statistician should take into account the total context of the study. This is where judgment, expertise, and plain old opinion enter the science of statistical analysis—something to keep in mind when reading that a favorite dessert now causes cancer.

That basically concludes the story of the Stanford Heart Transplant Data: No one was able to make any real sense of it. It does serve the purpose of amply demonstrating the vagaries of statistical research.

Thus far, we have talked about and illustrated the data and analytical problems previously identified for this dataset. Now I wish to divert our attention and show that much of the ambiguity can be removed, clearing the way for more definitive statements. To do this I will use the methods illustrated in lesson 6, changing sample sizes, and expand on them a little. Recall that most researchers would not do this, even if they thought about it, because it violates the rules of analysis. Remember that sample size is sacred.

We can cast more light on the variability encountered here by considering the starting times of admission to the study, and the associated waiting intervals. We argued that there is nothing sacred about that admission time—it is arbitrary.

According to the initial report, patients were admitted to the transplant study only after extensive questioning and testing, a meeting of the medical experts, and what seems to be a matter of "time to get around to it." For example, when all the testing and interviewing was completed, the patient was classed as "first being considered." Some time later, if all went well, the patient received "acceptance" and the clock was started. The time between these two events varied from one day to several months. Was the patient in some kind of health-stabilized holding pattern during that time?

What was sacred about that "acceptance" time? Why not use the "first being considered" date?

Let's see what might have happened if the additional quasi-waiting time between being considered and being accepted had been counted. (We won't even consider the years that most of these people had been seriously ill.) To simulate that scenario, we added a random number of days, from zero to sixty, to precede the acceptance date. A different number is used for each subject. All other dates stayed the same; these additional days were added at the outset. We ran the problem for fifty samples of added random times while considering the possible changes in relative risks for age, date (of admission), prior surgery, and the transplant. This is equivalent to running the same study, with the same people, fifty times over. But each time each patient is "admitted" on a different date. (I refer to this changing of acceptance times as "jittering" the start dates.) Consider it as fifty different tests or samples.

The resulting ranges of risks are shown by the heights of the heavy black bars in figure 7.8. What the picture says is that the chances of surviving, as they may depend on the first three variables, age, year, and PS, don't really change in relation to when the clock starts. But that all-important one, transplant operation, moves all over the place. Well of course it does, I'm changing the waiting interval and all the "measured" survival times of the nontransplant cases. This little game also mixes up the sequence of deaths and censorings as it appears in the computer, because the model lines up all the starting dates at zero to begin the calculation. The important dates then get pushed around with respect to each other. That's not cheating, it's the way things would have worked if the subjects had been accepted on those imagined dates, even if they did die or leave at the same times as before.

So we see that the critic's concern over the waiting interval usage was justified, but he apparently recognized only a tiny portion of the real problem: The so-called survival times were measured with a rubber band. Determining survival time with a bad heart is not unlike stating how long your last cold really lasted, down to the minute! Would you start the clock at the instant of your first cough, or perhaps when you took the first cold tablet, or how about when you dialed to call in sick? As we have said, some planning and discussion before the start of the study might have alleviated the problem. This exercise simply demonstrates how carried away people can get. It also highlights, once more, the need for real planning.

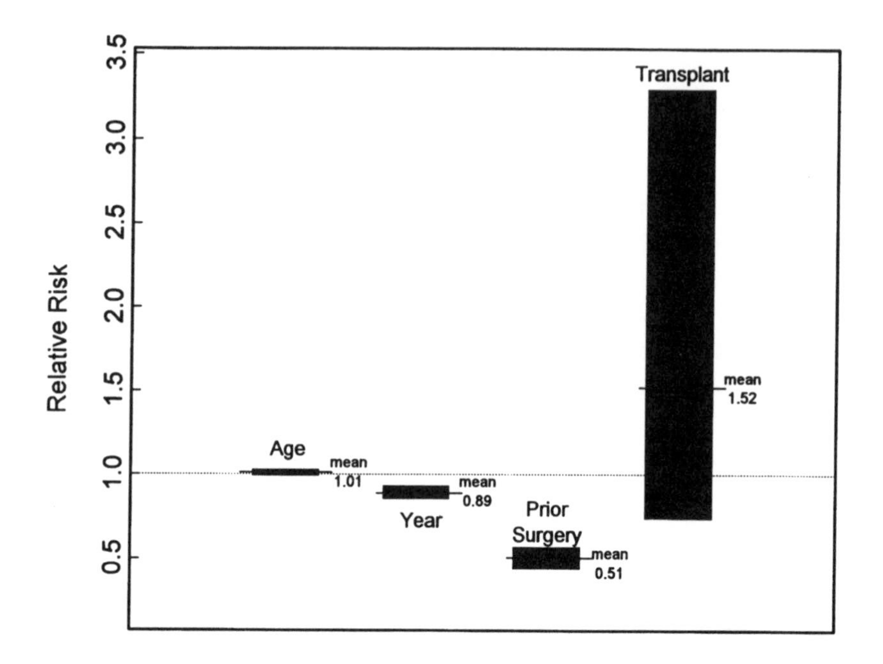

Figure 7.8: Ranges of Relative Risks Obtained by Jittering Start Dates

---

On the other hand, why aren't the effects of the other variables significantly changed by the jittering? Those thin little black bars are practically just lines; the risk for age, for example, hardly moves from the mean value of 1.01. First of all, these are not interaction effects. The model uncouples the effects of transplant from each of the other variables. In essence, it includes transplants and non-transplants all in the same pot when it considers age or prior surgery. Any uncertainty associated with the transplant is lumped into the last tall black bar. What the picture clearly states is that the estimated risk factors associated with age, etc., for people with these very bad hearts are as given by the mean values shown by each bar. They don't change whether the subject receives a new heart or not. This presentation has little of the ambiguity associated with the one sample analyses presented in the preceding tables. It is clear that the first three variables are consistent but the transplant numbers are highly questionable.

In the series of tables above, one of the sets of results was for analyses done part way through the study, eighty-two patients instead of the full 103. We mentioned that differences noticed by authors of the later report were written off as probably due to the different sample sizes or amounts of data. Although the reasons for doing an analysis part way through the work could have been many, that is not important; the choice of a point in time or the number of subjects to include is just another arbitrary item. Picking up on that, we can ask, what if analyses had been done regularly, say every sixty days for example?

That's easy. We'll go through the data in sixty-day increments, cull the subjects who were included up to the point, modify their survival or waiting times and dispositions to be consistent with that new observation point, and run a hazard model. We did this beginning at about two years (21.5 months) into the study, and considered the estimated relative risks for a number of the variables. The results are illuminating and are shown in the next several figures. Here again we can easily sort the good numbers from the bad.

Figure 7.9 shows the operation itself, the transplant variable, as it corresponds to our table 7.1 (p. 186). The value 1.08 in table 7.1 labeled "Mine" appears as the last point on the plot. On the plot we also show the approximate time of the published interim report on eighty-two subjects with the vertical dashed line, and the null relative risk value of 1.0 (no effect) by the horizontal dashed line. Remember, points above that line mean the transplant is bad for you, points below mean it is beneficial. Now, what relative risk value would we like? Pick one and then we will assume that the study terminated at that point. It is obvious that the analysts who chose to believe the results from the full dataset over the results from the partial were indulging in wishful thinking at best. Figure 7.9 is yet another example of why studies conflict: Each point here could easily be taken as the result of a study by another group. Sample sizes are always different in different experiments. With that in mind, it is clear that results of different experiments must be equivocal. That's a pleasant way of saying that it is very difficult to replicate experimental conditions, particularly when the studies are ill-planned or misguided. If we could plot all the results of the alcohol-heart disease studies that have been reported in recent years, it would look like this. When our risk factor results look like a random scatter diagram, we had better go back to the drawing board. The only way we can be comfortable with statistical studies is when they present consistent results. Without consistency, something is wrong. We should always remember this when seeing the really exciting headlines broadcasting the

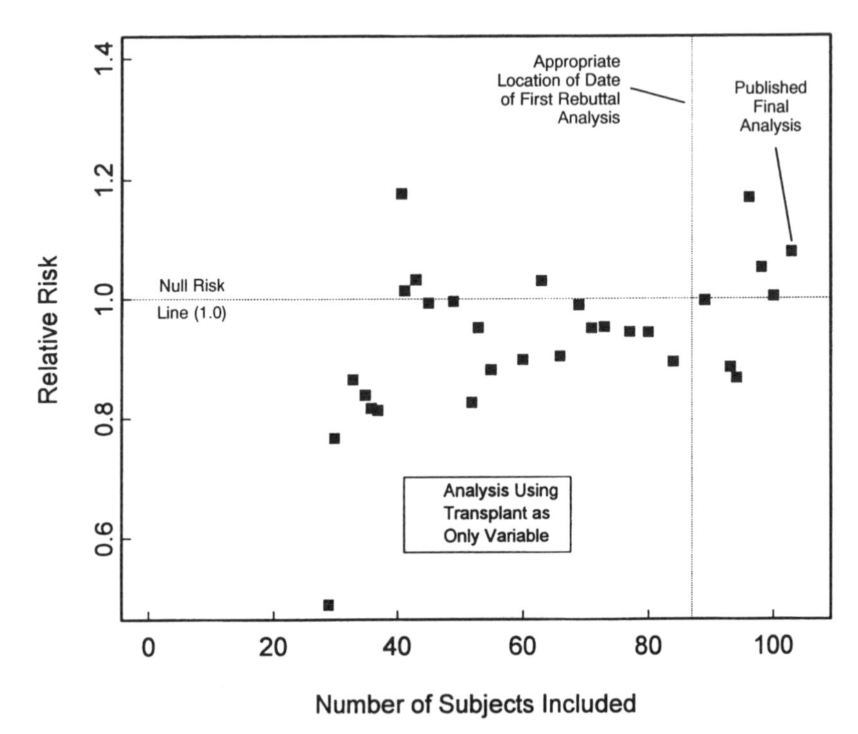

Figure 7.9: Analyses Every 60 Days Beginning 21.5 Months Following First Admission

results of "a" study. How consistent can one study be, and with what? By the way, it seems that as time goes on, more and more health columns are creating headlines out of single studies.

A few of the many recent examples include the following: in 1994, "Childbirth, Cancer Link Revised" and "Cholesterol Drug Saves Lives," in 1995 "Study: Tofu Is Latest Cholesterol Buster." Each of these reports was based on a single study. Now that almost no one worries about samples any more, are we being pressed into yet another level of conditioning?

In table 7.2 (see p. 187), we showed that when age at admission was included in the analysis, the risk of transplantation switched from being bad to being good. Figure 7.10 shows the sixty-day period results for that combination of variables. With age in the picture, transplantation turns up almost always being something good (there are only a couple of bad risk values), but the range of risk values is still gross and inconsistent. Not only that, as the number of

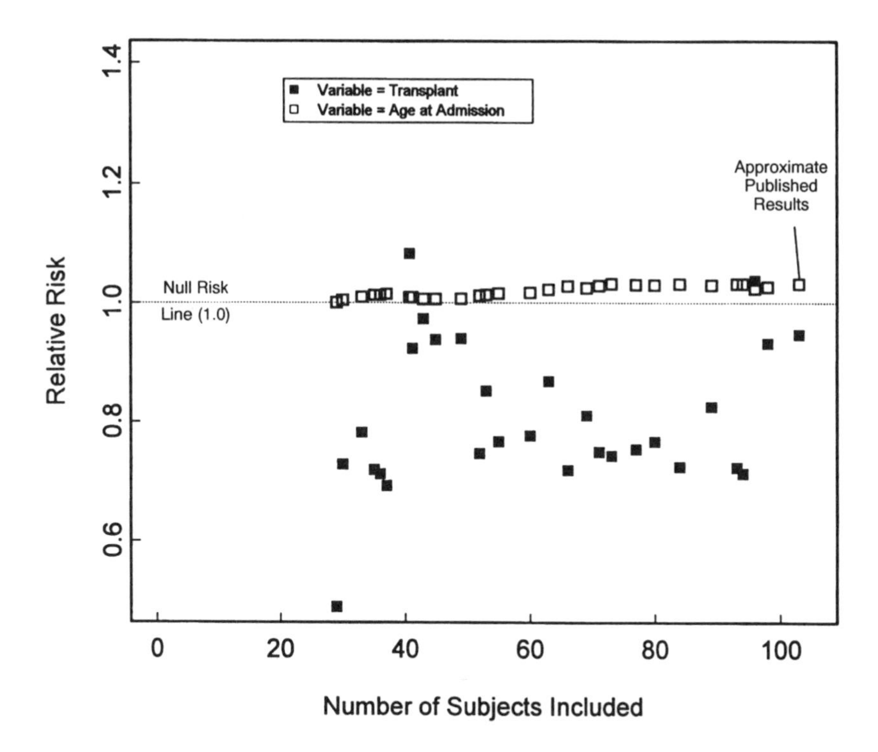

Figure 7.10: Analyses Every 60 Days Beginning 21.5 Months Following First Admission

---

subjects increases, there is no tendency for the variations to decrease. There is no trend toward a final value as there should be if only sample size were the problem. This scatter is more evidence that the study or the analysis or both are seriously flawed. Notice however, the variable age is nearly constant at a value near the 1.03 number of table 7.2 no matter when the analysis is done. This plot is then another comparison of stable and unstable results, which translates into believable and not believable.

Now consider the waiting time. In table 7.3 (p. 188) we claimed that when that variable was tossed in the ring, neither the length of the waiting interval nor the operation itself had any bearing on survivability. Figure 7.11 shows that the result for the waiting time is essentially unchanged by doing the analyses at different points in the study, but our old friend transplant is all over the lot again. The double null (the two values of 1.00) in the last column of table 7.3 was a sheer coincidence.

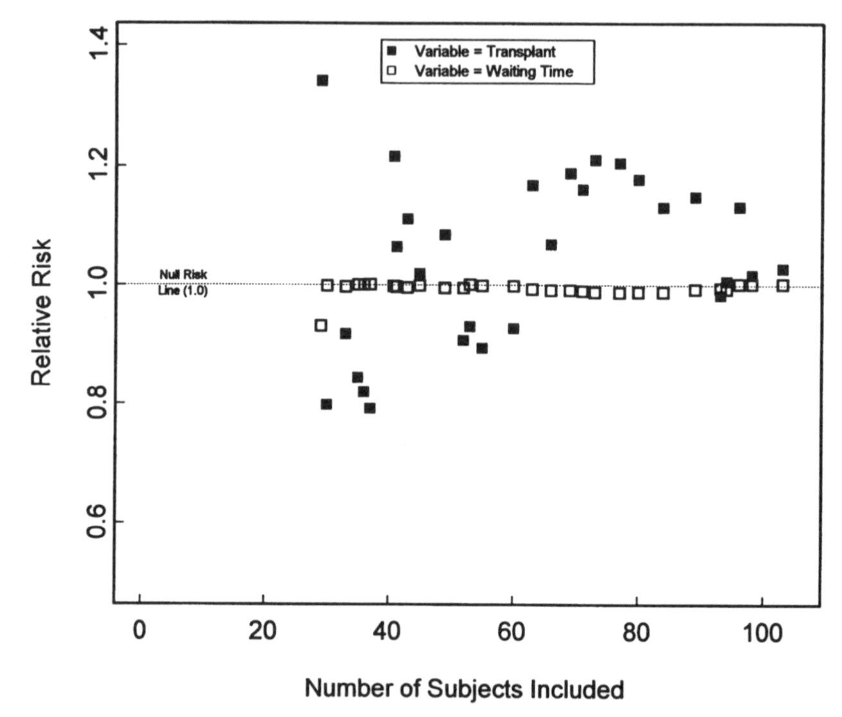

Figure 7.11: Analyses Every 60 Days Beginning 21.5 Months Following First Admission

---

Remember, in all these figures, only the last point of each plot, the values for the complete study, are analogous to the ones that were published in articles covering the full period. Those "final" numbers, when they became available, took preference in believability over the few interim ones, simply because the researchers had more data to work with. Those numbers, based on *all* the data, must have been the correct ones. No one bothered to wonder about consistency of result versus sample size or length of study. No one asked if the results would change if there had been fewer patients. That is not a recognized statistical question. And so it is with all studies, only the final-result values appear, unquestioned, in today's health columns.

By now some readers may be a trifle bewildered by the consistent inconsistency of the results. It seems that whenever we look at a new variable, all prior conclusions reverse themselves. I have rather casually mentioned that researchers must sort things out.

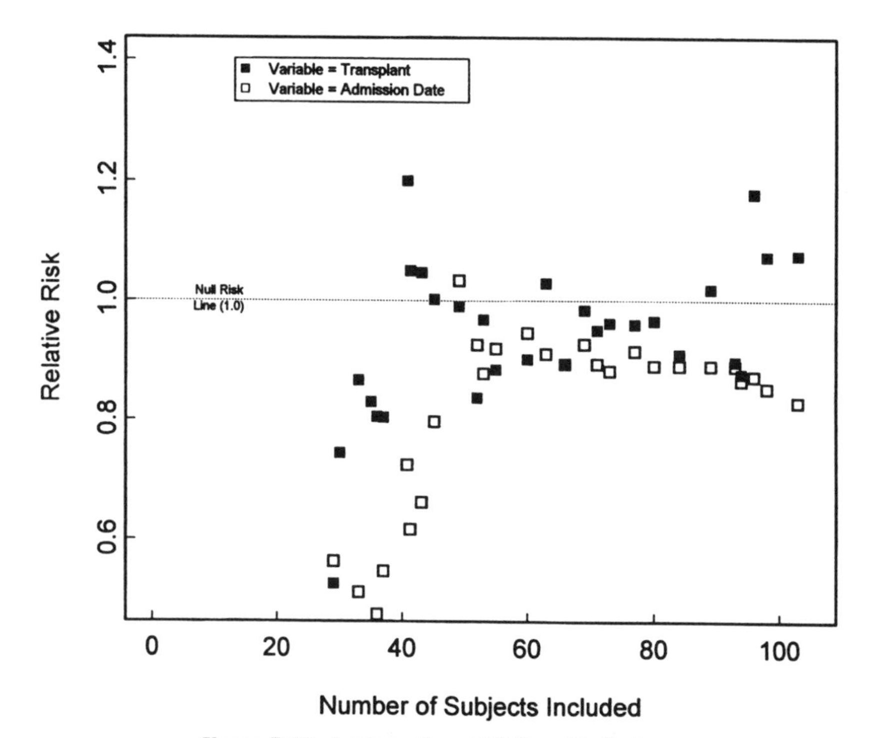

Figure 7.12: Analyses Every 60 Days Beginning 21.5 Months Following First Admission

---

That's true, but it is often easier said than done. The process is filled with trial and error, postulating and testing explanations, and so on.

It turns out that this technique of creating numerous samples, or doing periodic analyses, sometimes eases the process of sorting the real risks from apparent ones. A quick example will illustrate.

Consider the date of admission again. From our preliminary look it seemed that things improved as time went on. That is, survival periods seemed to be longer for patients entering the study at later dates. This was also a conclusion, though hedged, by many of the analysts.

Figure 7.12 shows the sixty-day periodic analysis relating transplant and starting date, which are considered as separate factors. Again, the last open and solid squares on the right correspond to the published results. They indicate that those who underwent transplantation had a higher risk of death, but those who started (entered) the program late had a lower risk of dying. As we expect by now, the value for transplant risk is all over the lot again but the admission date is somewhat better behaved. At least it (essentially) stays below the null risk line. The rising values on the left side are probably due just to the sample size becoming large enough for the effect to stabilize. In view of that, one might expect that the later risk values for starting date, say after about fifty or sixty subjects, should be more reliable. Oddly enough however, the only acceptable P-values (not shown), those less than 10 percent, appeared at the two extremes of the plot, that is, for sample sizes of less than forty subjects or more than about ninety. Thus the P-value measure says we should believe either the very small values of risk or only those near the end of the study. Incongruous isn't it? In spite of this strange behavior of the P-value measure, which is not unusual, the researchers consistently hold this up as *the* measure of how reliable their study is.

Now, as a second try at making sense of the effect of start date, we can look for any interaction or combined effects of date and transplantation somehow working in concert. For example, suppose it were true that the surgeons and the rest of the medical staff were becoming ever more expert at transplantation and the care of transplanted patients. Further assume that the care for nontransplanted patients was unchanged. This would appear as an interaction because patients admitted later and receiving a new heart would benefit doubly, but no change should be found in the survival of nontransplanted patients. Interactions of this nature are often detectable in a model by simply multiplying the numbers for date by the transplant variable (a one for transplanting and a zero for not.) This new artificial product is called an interaction term and then used in the model just like any other variable. The result is shown in figure 7.13, and everything seems to blow apart. Both variables now fall all over the place. If you look closely, you might notice that the relative risks are traded back and forth as the sample increases. That is, in the first part of the plot, for smaller samples, if a black square falls above the null risk line, then its corresponding white square is below it, and vice-versa. It's as though the model can't make up its mind as to which one to blame. Note that in the prior case, figure 7.12, the black squares were almost always above the white ones. Don't fret though, that change in relative positioning of the squares from figure 7.12 to 7.13 may be an indication that something significant is buried here, we just haven't discovered the right shovel. More work is needed.

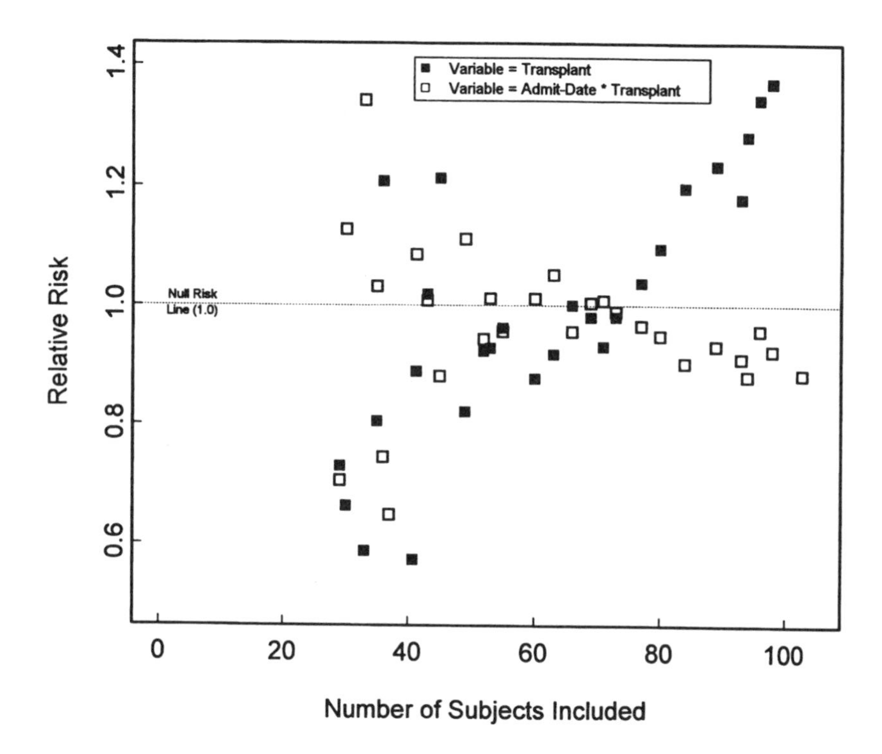

Figure 7.13: Analyses Every 60 Days Beginning 21.5 Months Following First Admission

---

So far we have tried the date and the date-transplant terms as separate variables in conjunction with transplant. In both cases the results were a good indication that we are still doing something wrong. You might interpret this as meaning that the total risk is not nicely divisible among those pairs of variables and something else must be involved. A next step is to try all three together.

This result is shown in figure 7.14. Now, for the first time we have a picture in which all the variables considered, even transplant, are reasonably consistent. They start to settle down once the sample size exceeds forty or so. It appears now that both having a late starting date and undergoing a transplant make for smaller risks and that the interaction term is soaking up all the bad aspects, i.e., risks greater than 1.0. However, it is not easy to know just what that means. It could be that the interaction term is simply acting as a surrogate for some other factor not included in this run. This is much like researchers saying that there is something effecting the results

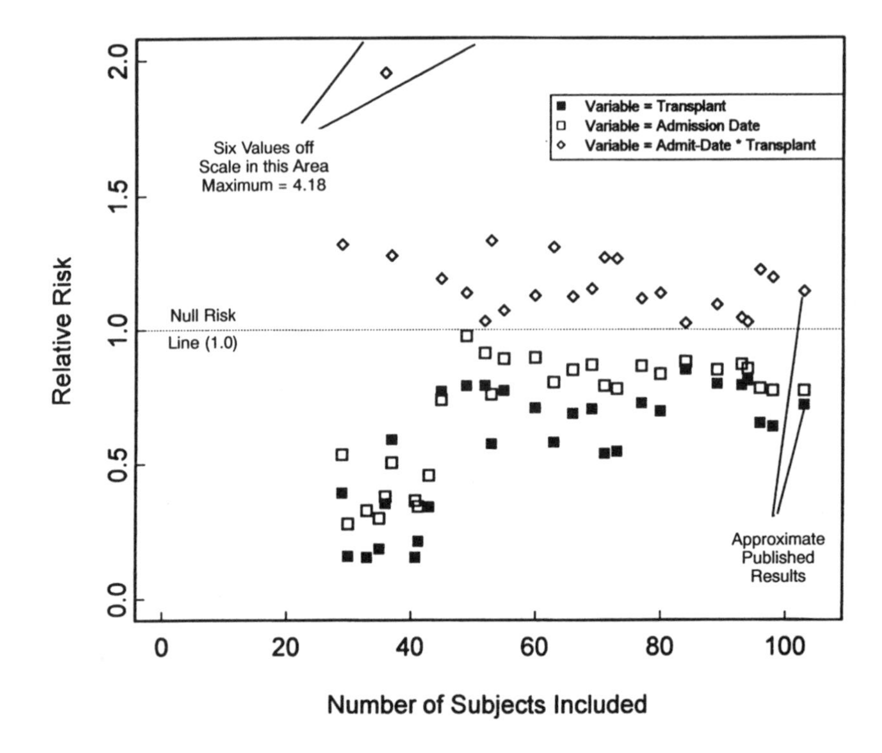

Figure 7.14: Analyses Every 60 Days Beginning 21.5 Months Following First Admission

---

but they don't yet know what it is. At least it seems that we are on the right track. We should be encouraged to explore more.

Figures 7.12 to 7.14 illustrate the trial and error analyses that must be conducted to sort out and identify conflicting results. These particular plots make it relatively easy. Pity the poor researchers who don't examine their data like this. They blindly pour their data into the computer, make variations of the variables until "good" P-values fall out and proclaim "voilà, those other studies were wrong."

Now we are prepared to appreciate how including the waiting time intervals in the manner of the published study reports really messed up the picture. In figures 7.9 through 7.14 we saw how the estimated relative risks for a number of variables fluctuated wildly depending on just when the model was run (sample size or number of subjects). If we do the same sort of sixty-day interval modeling for the data but this time ignore the waiting time we get very different

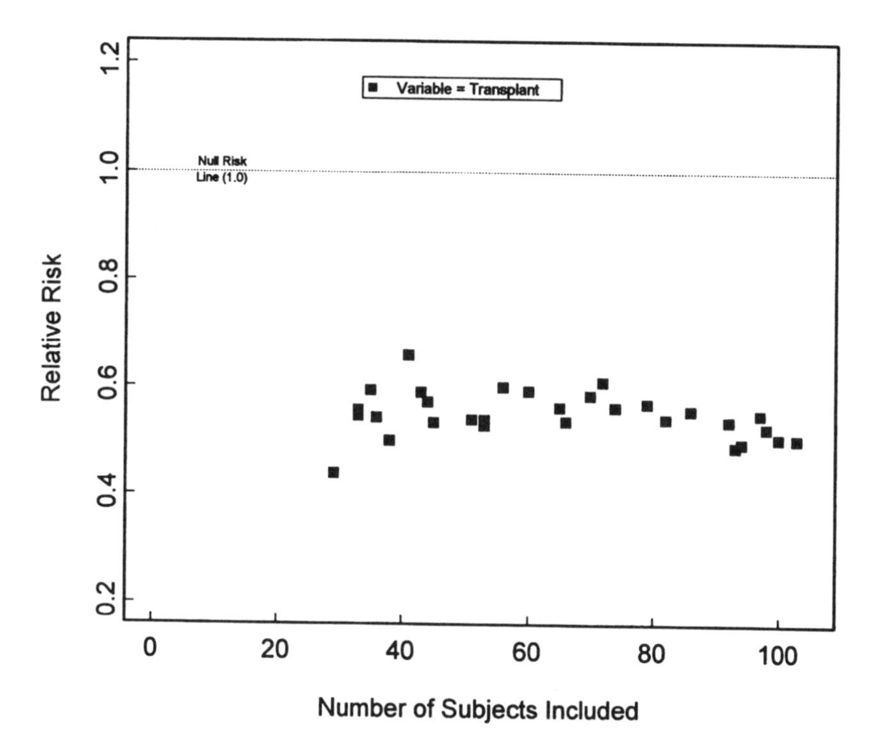

Figure 7.15: Analyses Every 60 Days Beginning 21.5 Months. No Waiting Times Included.

---

results. It is important to understand that the pictures to be shown are quantitatively no more valid than the previous ones. There is still no real control group for the transplant cases, and starting the clock at admission date for the others is still arbitrary. My contention is that tossing in the waiting times amounted to simply adding a set of random numbers to an already bad dataset, and none of the numerous come-lately analysts realized it.

There is no need to perform makeovers of the previous six plots. All the variables however, even those not shown earlier, react to the inclusion or exclusion of the waiting time in similar ways. Briefly, they lose any semblance of sense or order when faced with those random waiting periods.

Figure 7.15 shows the relative risk for the single variable transplant, as estimated in sixty-day increments. The scale for risk is shifted from that of figure 7.9 from 0.6 to 1.4 down to 0.2 to 1.2 to accommodate the smaller values, but the range of the scale is the

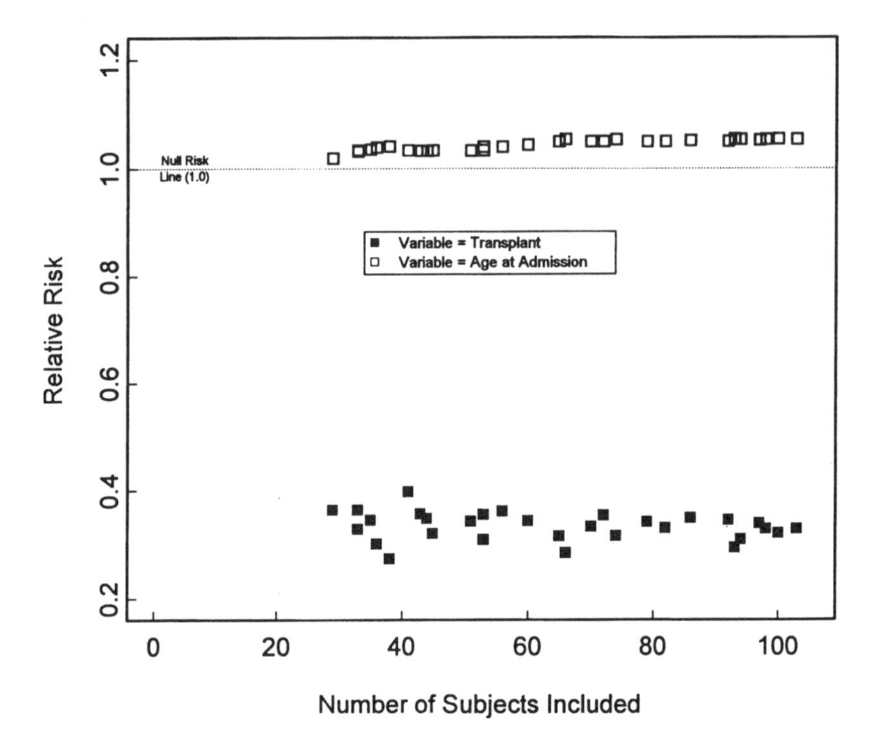

Figure 7.16: Analyses Every 60 Days Beginning 21.5 Months Following First Admission. No Waiting Times Included.

---

same, so visual comparisons are fair. See how relatively well-behaved or consistent the risk becomes when waiting time is removed?

Figure 7.16 compares with figure 7.10 which shows the variables age and transplant. In figure 7.16 the risk for age has moved up very slightly; the risk from the operation has decreased a lot and has become much more stable.

Looking at the next figure, 7.17, and its counterpart for date and transplant, figure 7.12, notice that the two are totally different. The date variable shrinks to the no-risk value, 1.0, and doesn't move, a far cry from the scatter in figure 7.12. Again, transplantation risk has calmed down.

If the waiting time had not been treated as it was in the actual study reports, and well-behaved results similar to those in the last few figures had been obtained, one might have considered the analysis to be complete, but this is not so, all possibilities must be looked at. One of them now of course is that product interaction

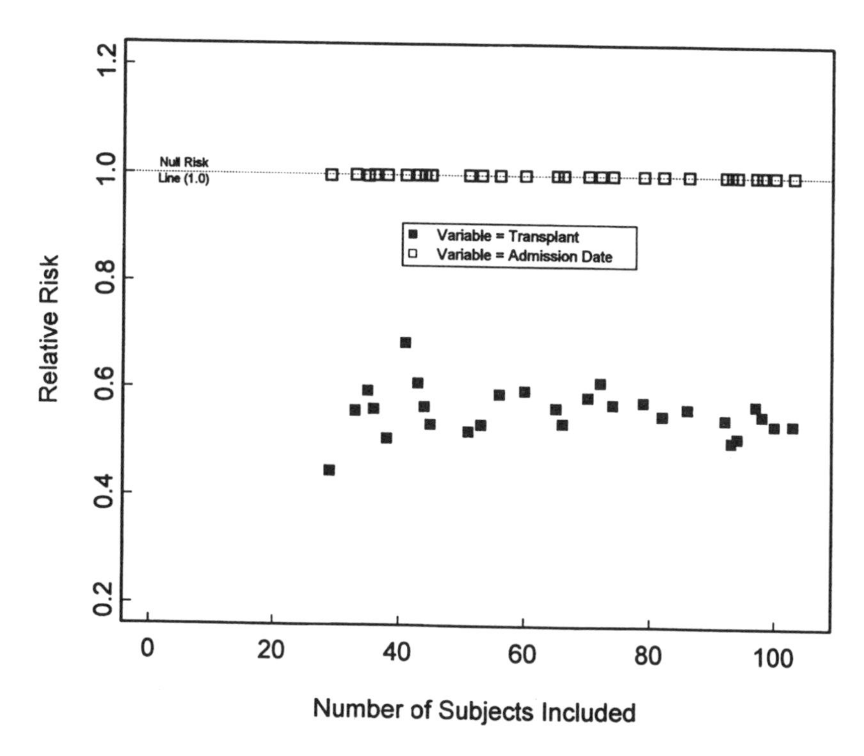

Figure 7.17: Analyses Every 60 Days Beginning 21.5 Months. No Waiting Times Included.

---

term of transplant and admission date. All such terms must be examined, there is no way the researcher can rule out any of them without first testing for effects. So, when we examine the three-way run with transplant, date, and their interaction, the changes from figure 7.14 to 7.18 are dramatic and obvious. However, compared to the date-only run with transplants, figure 7.17, the variability of the transplant risk is greatly increased. It seems that the model is becoming confused by the addition of this third variable, the product term. In trying to fit an additional parameter to the same data, it loses something. This behavior indicates that the new variable should be tossed out. With hindsight now, we suspect that the poor risk shown by the interaction term in figure 7.14 was compensation for the noise (superfluous data) introduced by the waiting time.

These figures show vividly why I said that the waiting time did nothing but obfuscate the results.

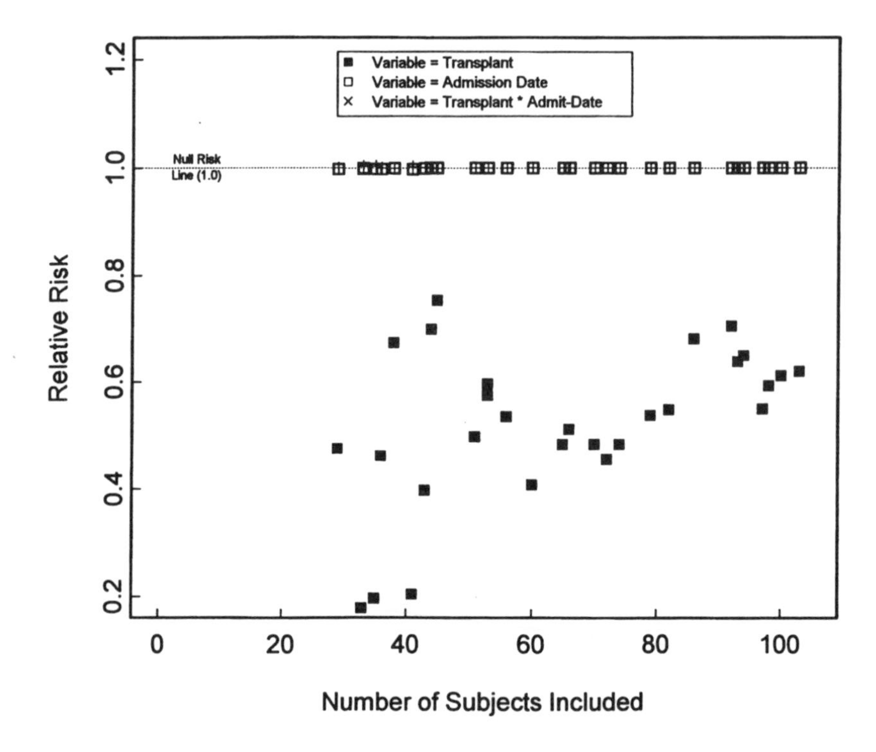

Figure 7.18: Analyses Every 60 Days Beginning 21.5 Months. No Waiting Times Included.

A few pages back, when we showed figure 7.2, illustrating the disposition and survival times of subjects, the question came up regarding a bias toward those who entered the study near the end. Possibly the researchers weren't given enough time to determine how long these patients might have survived. I promised to explain this, and now is the time. Figure 7.19 shows what are called "survivor functions," for transplanted and nontransplanted patients. The solid staircase-like line is for transplanted patients, the broken staircase-like one is for the nontransplants. We will get to the slanted dotted line later.

The bottom axis is labeled "Time," referring to days into a kind of idealized study in which all subjects enter on day one. The time is then days into the study but also survival times, as explained below. The vertical dimension, "Surviving Portion," refers to the fraction of all the subjects considered that are alive and still in the study at the specified day. Thus, at the first day (1), the fraction 1,

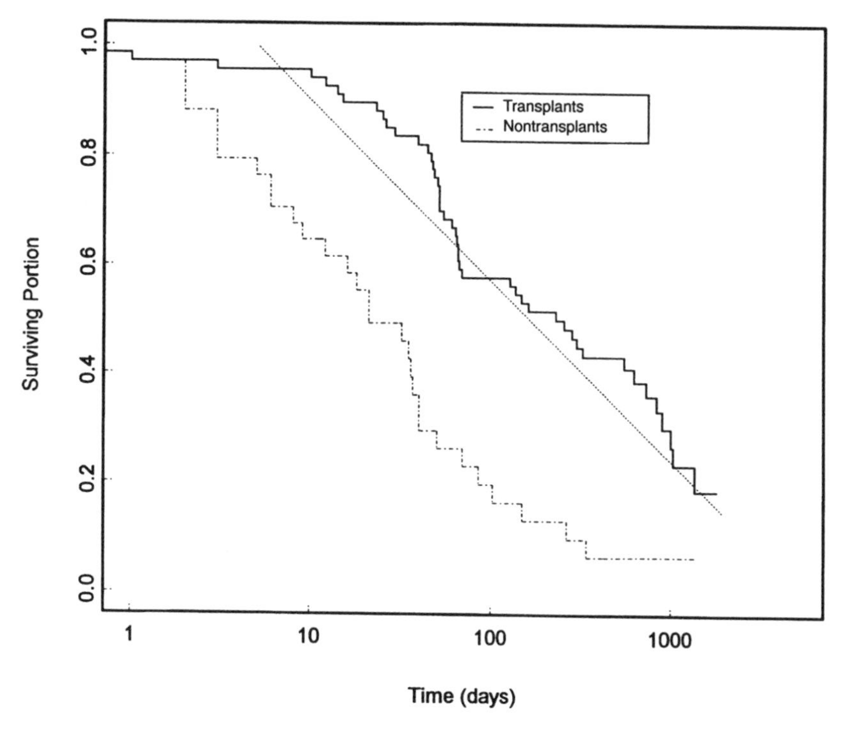

Figure 7.19: Survivor Functions for All 103 Subjects

---

or 100 percent, of all the transplanted patients were alive. As time progresses, members drop out by dying or through censoring for some reason, and the fraction of survivors decreases. Eventually it would drop to zero when everyone was dead, but the study might end before that, as it does here.

Each point on the curve, that is, the upper edge of each stair, is the fraction of the whole group still alive and in the study at that time. It is sometimes calculated in a special way, however. Roughly, the number that die on say, day 15, is subtracted from the total number that start out on that day, and the result is divided by the total number. Big deal you say, that's what any fraction is. Well, it is not always that clear here because of censoring. Remember censored subjects don't die, they just leave the study, and the question is how to account for them. They were included in the calculations up until the time of censoring, will not be included in future ones and yet do not contribute to the portion that die.

Well, suppose two people are censored on some day. Then nobody dies, there is nothing to subtract, and the fraction is 1. So this "new" point accounting for censored subjects is exactly the same as the previous point, it does not show up in the plot because there is no change in the height of the stair step. But, now the total number available to make the next calculation at a death event is smaller by two. So, between two steps, the divisor has changed by more than the number that died. That is perhaps a subtle distinction, but it is a neat way of handling the censored folk. Each point on the curve is still a best estimate of the probability of dying, based on the actual number of subjects on hand on that day, regardless of how many may be censored. It's considering the censoring to have occurred at the end of the day, after any deaths for that day that may have taken place. In general, when there is censoring, this makes the following step a little higher than it would be if no one had been censored, because of the smaller divisor. What it comes down to is that the censored folk are available to make the probability estimate when they are in the study but then they just evaporate, if you will, and have no further influence on the arithmetic.

Now, the hazard model handles censored subjects in much the same way as the survivor function plots, so those people who entered the study near the end and show up as the little open squares at the top right of figure 7.2 don't represent any bias. In essence, they made their contributions to the guesses of the probability of dying even though they were in the study only a short time.

By the way, the reasoning behind including the waiting time information as censored nontransplant survivor time is similar to this argument. Up until the time of transplant, those people were available for calculations of the chances of nontransplant death. That would have been perfectly okay, and even highly desirable, if the so-called starting times had had any meaning.

We have harped on the fact that this heart transplant study was not a study at all in the statistical sense. That is true, but please don't get the idea that all the problems, inconsistencies, and contradictions are because of this fact. More often than not, real studies have similar problems. This example was picked not only because it displays all these problems, but also because it shows how the best of experts can get carried away and ignore the obvious, in addition to failing to check their data.

Let's reflect back now on the many scatter plots in which we changed the sample sizes or times of analyses. There are some helpful hints to be picked up from these experiences. In nearly every case, the transplant variable was highly erratic. For the other variables, there was at least one plot, usually more, in which the associated risk values were reasonably stable and consistent. This tells us some things about using hazard models, as well as others that include many parameters.

Never settle for the first few obvious model runs. Make them for every conceivable combination of variables and interaction terms. Vary sample sizes or other parameters suitable to your dataset, and watch for instability or wildly fluctuating results. These should be discarded. The only results worth considering are those displaying consistent behavior. When results bounce around it may indicate one of many unpleasant things. Here are just a few:

- A bad sample.
- Poor experimental procedures.
- The wrong variables are being used.
- There are unknown variables that are the real effectors in the problem being considered.
- The wrong model is being used.

Researchers are usually educated about warnings of this type, but we have just been through a classic example of how easily the danger signs are ignored.

It is extremely difficult to determine that such problems exist without performing the sorts of functions we did here. People wind up doing all manner of alternative tests and modeling, writing paper after paper, even inventing new techniques, when the real problem is just another useless heart transplant study.

Just because many studies have been flawed does not mean that all is hopeless. In the next lesson we will discuss a good study.

Notes

1. Bruce W. Turnbull, Byron W. Brown Jr., and Marie Hu, "Survivorship Analysis of Heart Transplant Data," *Journal of the American Statistical Association* 69 (March 1974): 74–80.

2. David A. Clark et al., "Cardiac Transplantation in Man, VI. Prog-

nosis of Patients Selected for Cardiac Transplantation," *Annals of Internal Medicine* 75, no. 1 (July 1971): 15–21.

3. Mitchell H. Gail, "Does Cardiac Transplantation Prolong Life?" *Annals of Internal Medicine* 76, no. 5 (May 1972): 815–17.

4. John D. Kalbfleisch and Ross L. Prentice, *The Statistical Analysis of Failure Time Data* (New York: John Wiley & Sons, 1980), Appendix 1, dataset III, pp. 230–32; and John Crowley and Marie Hu, "Covariance Analysis of Heart Transplant Data," *Journal of the American Statistical Association* 72, no. 357 (March 1977): 27–36.

5. Statlib, On-Line Library for S-Plus Software Users, Maintained by Carnegie Mellon University: <statlib@lib.stat.cmu.edu>.

6. Turnbull, Brown, and Hu, "Survivorship Analysis."

7. Byron W. Brown, Myles Hollander, and Ramesh M. Korwar, "Nonparametric Tests for Independence for Censored Data, with Applications to Heart Transplant Studies," in *Reliability and Biometry: Statistical Analysis of Lifelength*, Frank Proschan and R. J. Serfling, eds. (Philadelphia: Society for Industrial and Applied Mathematics, 1974), pp. 327–54.

8. John Crowley, "Asymptotic Normality of a New Nonparametric Statistic for Use in Organ Transplant Studies," *Journal of the American Statistical Association* 69 (December 1974): 1006–11.

9. Crowley and Hu, "Covariance Analysis."

10. Kalbfleisch and Prentice, *The Statistical Analysis.*

Lesson 8

They're Not All Bad

At this point, some readers may have the feeling that there is no hope of ever learning anything worthwhile from statistical studies, particularly in the world of medical subjects, in spite of the tiny words of encouragement at the close of the last lesson.

Well, not everything is lost. However, finding good studies is not easy for an outsider. As I mentioned early on, it is generally difficult to access and examine raw data files to find one.

After some effort I had to resort to one available in the database supplied with some software known as S-Plus[1] (a very nice package). It has to do with a problem associated with spinal column surgery in very young children. It is not clear why the surgery is performed, but it can involve any number of vertebrae along any part of the backbone. It turns out that sometimes there is an undesired result or side effect called kyphosis, a fancy word for humpback. (It may not always be as bad as one might think, apparently there are degrees of humpedness.) The object of the study was to see if any conditions or circumstances could be identified that might lead to safer procedures for the operation.

The data are quite simple: the age of the child in months (A); the starting location (S), identified as the number of the first vertebra involved; and the number (N) of vertebrae operated on. There are thirty-three distinct vertebrae in a child (as opposed to thirty-two in an adult), and convention labels the topmost one as number 1. So, the starter (S) may be any number from 1 to 33 and the number involved (N) must range from 1 to something less than 33. With each set of data, A-S-N, there is a number, 1 or 0, indicating whether or not kyphosis occurred. There were eighty-one patients and seventeen developed the problem. This gives us a reference estimate of the chance of getting a hump, 17/81 = 0.21 or about 20 percent. We won't concern ourselves with errors or confidence bounds for that value, because (surprise) more work is needed, as you will see.

Following my own advice, the first thing we do is look at the data. This is really easy with such a small data base. The three scatter diagrams in figure 8.1 make a good start. Plots for the three variables, age, start, and number, show their distributions among the eighty-one children. The occurrence of kyphosis is shown where the asterisk is replaced with the letter K. I also noted on each plot the correlation between the variable and the occurrence of a hump.

If we study the three pictures for a while, it is clear that there are mild trends of the K points with the variables. They are visually evident and the correlations support the observations. It is curious however that the Ks seem to thin out as one moves toward the lower or upper edges of the age scatter. Notice, however, that the visual trend of the relative densities of Ks in that scatter is not as evident as in the others, as confirmed by the coefficients, only 0.13 for age but –0.45 and 0.36 for the others. The trend of the Ks here is nonlinear, it decreases for both younger and older ages instead decreasing in only one direction (as the other two do), but the coefficient cannot detect that, because it "sees" only linear trends, it can't bend around corners. One suspects that the real relation between the Ks and age is greater than the coefficient is telling us. We will come back to that observation and make use of it a little later. For now, let's proceed with the initial analyses.

The example analysis in the source study looked for significant differences in the occurrence of kyphosis as measured by mid-values of variables, for those with and without the undesired event. This is a common choice of things to do early in the analysis, and it is also a linear one, that is, it uses straight lines to approximate all trends. We should always look at a number of these linear tools before jumping into the more complex material. As a rule, they are more rugged or stable, robust in the statistician's vocabulary, than the nonlinear ones. If satisfactory answers to the questions posed in a study can be obtained with the linear tools, they should be retained in the interest of simplicity. It is rare to have just cause, beyond the actual data, to impose nonlinear methods. (The table methods of lesson 6 are all linear.)

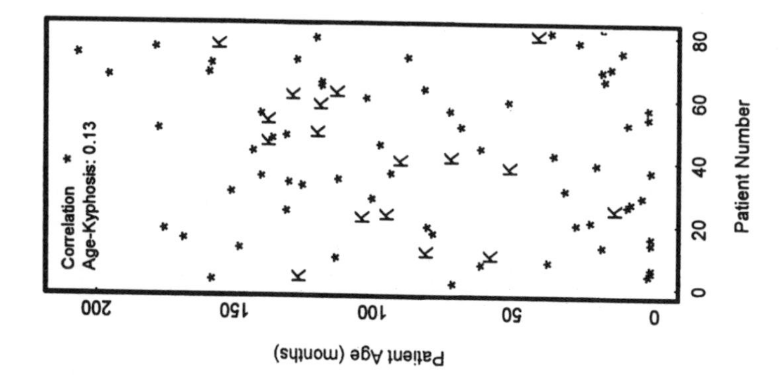

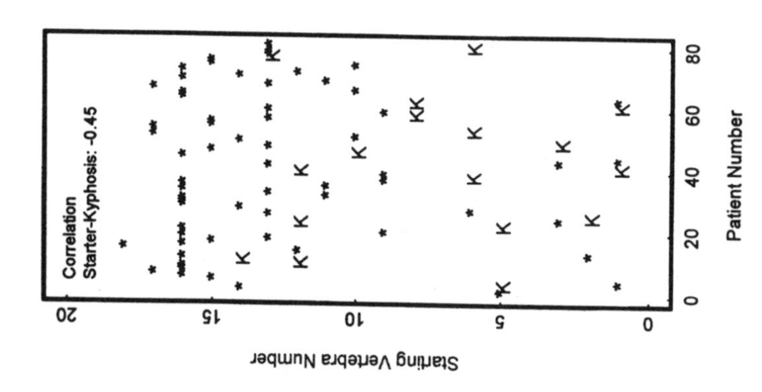

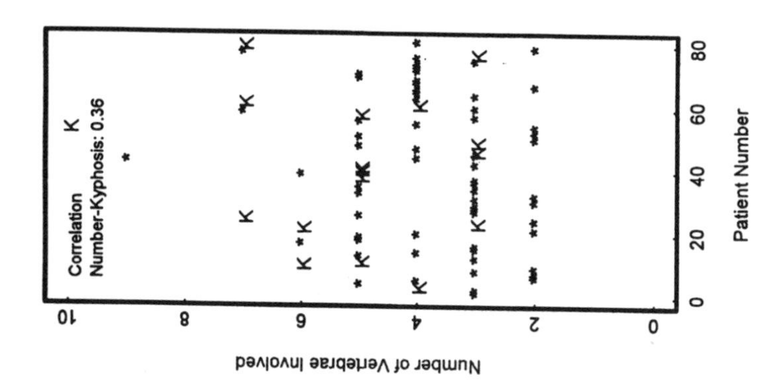

Figure 8.1: Starting Scatter Designs

In the following, we will extend the analysis beyond that considered in the source study to look at the variables in a number of ways, attempting to extract more information. We begin by first replicating the example to see the starting point. As mentioned above, the first look is to see if the average values of the variables are of any help in predicting kyphosis. The method of choice for the comparison of mean values is something called a boxplot and it needs some explanation.

In lesson 2 we imagined a pile of prone people stacked up to see a normal distribution of heights of Americans. When we stepped way back the outline was a bell-shaped line and we said that about 95 percent of all the folk were within plus or minus 2 sigma of the mean.* If we took a photo of all those feet, we could draw a box on it, surrounding the center of the stack, enclosing those 95 percent. Then suppose that we drew little lines, or whiskers, from the sides of that box out to near the ends of the pile, leaving about one-third of a percent outside the end of the whisker and then draw lines there, one near each end. Just for reference, we now draw a thick vertical line right at the middle of the box. What we have is a boxplot: a rectangle with whiskers and lines that summarizes the bell-shaped curve. Now look at figure 8.2. These are pinched boxplots. There are minor differences between these and the one we drew around the feet. The boxes include 50 percent of the data instead of plus and minus two sigma's worth, but the vertical whiskers do leave about a third of one percent outside, above and below, those horizontal markers. The thick line across the middle, the skinny part of the pinch, is the estimated median, or 50 percent-point of the distribution. The dark pinched area shows the confidence bounds on that median. That is, the true median should fall somewhere within that narrow band. If the distribution happens to be one of those nice normal ones we like so much, the boxplots would be perfectly symmetrical top to bottom. In this case, only one of them really is, that for "yes" in the third picture, which depicts the variable number (of vertebrae). Notice, in that picture, there are two little lines above the tops of the whisker limits. Those points were declared to be outliers by the software. They are more than two and three-quarters sigma removed from the median. The two and three-quarters is one of those arbitrary outlier criteria we talked

---

*Recall that sigma is the measure of dispersion or spread in things—the plus or minus five yards of the 2-wood drives.

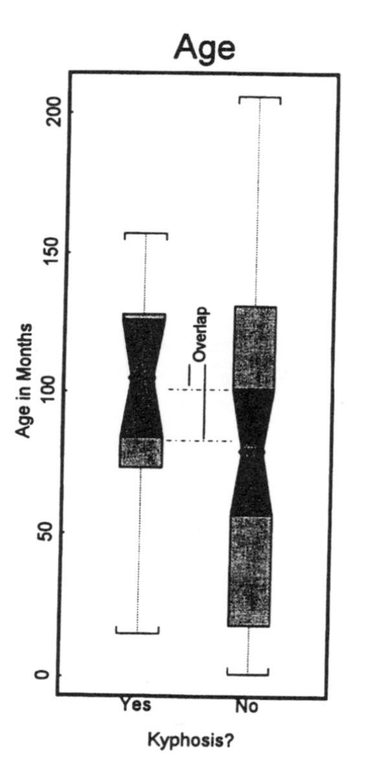

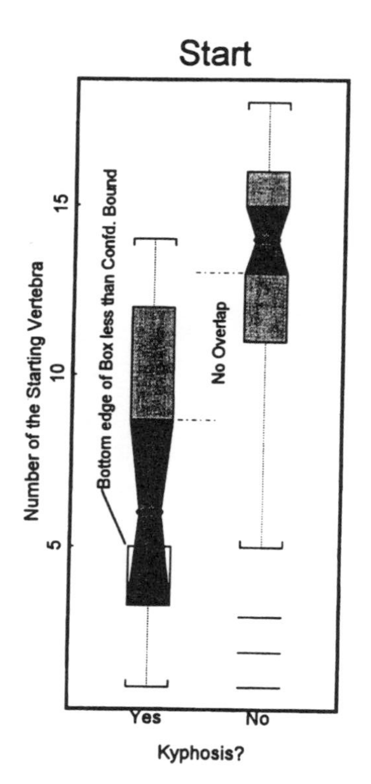

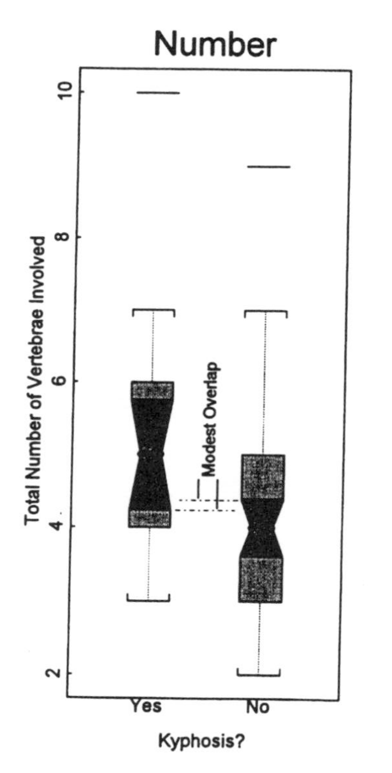

Figure 8.2: Boxplots Show Spread and Separation of Kyphosis-Tagged Variables in Months

about and happens to be quite useful here. The intent is just to call the outliers to our attention so we can take a closer look. Remember, they might be really interesting. Such lines might point out the shortest and tallest people in your photo of the pile of feet.

With these plots we actually compare median values rather than the means, but often that is adequate, because the means and medians are frequently very close to each other. In this case each box in a pair corresponds to whether kyphosis occurred and, if the medians are very different, we might have a clue as to the control of the undesired events. For example, in the first picture, if the "yes" box were a lot further above the "no" box than it is, we could say that only older patients were susceptible. A quick judgment as to whether the medians of two distributions are significantly different is made by using those pinched confidence bound areas on the boxplots. If the pinched regions in a pair overlap, then the difference is not significant. The confidence bounds of the pinched areas are such that they correspond to a significance level of 90 percent for differences of the medians. I put dotted lines on the pictures to point out any overlap. This first look indicates that only the starting vertebra is important in the kyphosis problem. Only the center pair of plots has a significant difference in the medians.

The three pictures in figure 8.2 correspond to the three scatter plots in figure 8.1, but are considerably more quantitative. Now we'll look at some other techniques that can extract even more information from this set of data.

Of course we can use the logistic model here to estimate the relative chance of getting a hump in terms of any, or all, or combinations, of the variables A-S-N. It is also possible to introduce an artificial time scale and use the hazard model. If is this done we should alter our choice of times, e.g., different sets of random times, and times ordered by patient number, to examine any effects on the answers. As expected, when this was done, differences were encountered.

We must remind the reader here that assignment of time slots to the occurrences of kyphosis has no meaningful interpretation. Unlike survival time of a heart transplant patient, it is not really sensible to think of time to development of the kyphosis problem. Within a limited time period following the operation, either the problem developed or it did not. All this artificial time assignment does is allow us to look at the data with a different model and vary the sequence of cases as well as the sample size. If the results are reasonably consistent we can feel more confident about the quality of the study and reliability of the results. Note that those terms, "quality" and "reliability," as used here are qualitative and subjective. No statistical test is implied. This should become more clear as we look at examples.

When hazard runs using random assignments of time slots were made with the three variables, A-S-N, the relative risks changed modestly and the associated measures of error, the P-values, moved significantly, especially for the variable A, age. Nevertheless, when the sample size was varied from a relatively small number (fifty-five) to the full eighty-one, the fluctuations were quite mild compared to what we saw in the last two lessons. In addition, once the sample size reached values in the vicinity of seventy, all the relative risks were converging. In other words, the patterns of dots appeared to be homing in on some final value for even the most highly varied cases. Also, the values toward which various runs seemed to be "settling in" differed by only small amounts, less than 10 percent. Recall that, with every change of the assigned times, the ordering of the sample was shuffled automatically, so we were not simply using the same group of last subjects in the runs and thereby achieving some sort of pseudo-consistency. Similar degrees of stability in the relative risks were realized when the logistic model, which does not concern itself with the ordering of events, was used to examine the same samples. We note again however, that even though the risk values were behaving in this consistent manner, the P-values, especially for the age variable, remained large, in the vicinity of 0.8 or 80 percent.

This prompted an even closer examination of those scatter diagrams of figure 8.1 to look for possible transformations that might assist us here. We saw earlier that kyphosis cases tended to favor the middle of the range of ages (as seen in the left plot of figure 8.1). It seems as though the chance of an event is greatest for the middle group, around 100 months of age, and tapers off for both younger and older subjects. The thought occurs that we might define a modified age number that is zero in the middle, goes negative for the younger subjects, and positive for the older subjects. Then, ignoring the sign, when this number is small, it represents an age group with a greater number of cases, and as the number gets larger, it progresses in the direction of fewer cases. In essence, this splits the group in the center and recombines it to make the direction of increasing cases of kyphosis consistent. We now have a linear relation between age and kyphosis with this modified age variable.

We can obtain such a linear relation as follows. If we subtract some number near the middle, say, the average age, from all the age values, then approximately, those from 100 months down will become negative, with values running from 0 to –100, and those above 100 will remain positive but have values ranging from 0 to +100. Now take the absolute value of the ages (make all the signs positive). This effectively folds the age scatter up over itself. The Ks are then most dense near the bottom edge, which is zero on this modified age scale, and thin out as age rises toward 100. The events now cluster at the smaller age values as we see in the left picture of figure 8.3. To emphasize the point, the center plot of this figure shows the same data but with the patients ordered according to their true ages. The Ks, most dense in the middle-aged group, are decidedly crowded toward the lower portion of the modified age scale. As hoped, the magnitude of the correlation has more than doubled, from 0.13 to –0.29. The value is now negative of course, because the chance of kyphosis goes down as this transformed age goes up. Using the modified age results in even greater stability with sample size in the regression runs. Before we look at that though, consider a second, possibly useful, modification.

Because some poor P-values were obtained in the analyses above, we should try some interaction terms among the variables to see if they might help. From figure 8.1 we note that both variables S, starting number, and N, number of vertebrae involved, have strong correlations with the kyphosis cases, though one is positive, the other negative. This says that both the location and extent of the operation are significant and suggests that an interaction term might be appropriate. Recall that interaction terms are used to evaluate joint effects of two variables and are obtained by multiplying them together. Before doing that though, let's change the counting order of the vertebrae to reverse the direction and cause the correlation to become positive. Then both increasing vertebra starter number and increasing number of vertebrae involved "work in the same direction" with their effect on kyphosis. We hope that the product of the two will then be very significant in the model.

So, in developing an interaction term we have created another modified variable, the reversed order starting numbers, as well as a new variable, the product term. A scatter of that product is shown in the right plot of figure 8.3, along with the correlation between it and the occurrence of kyphosis. The new coefficient is +0.48, and the trend, as seen in the plot, is now much stronger than in either of the individual scatters, the center and right plots of figure 8.1.

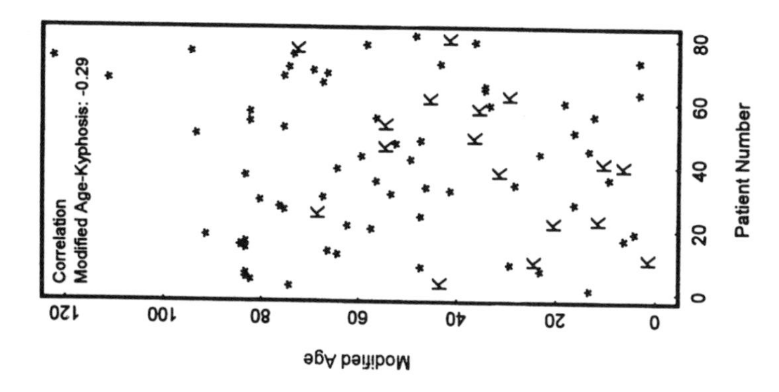

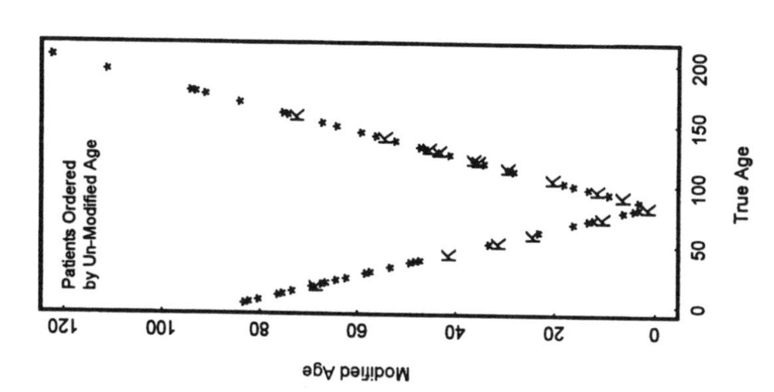

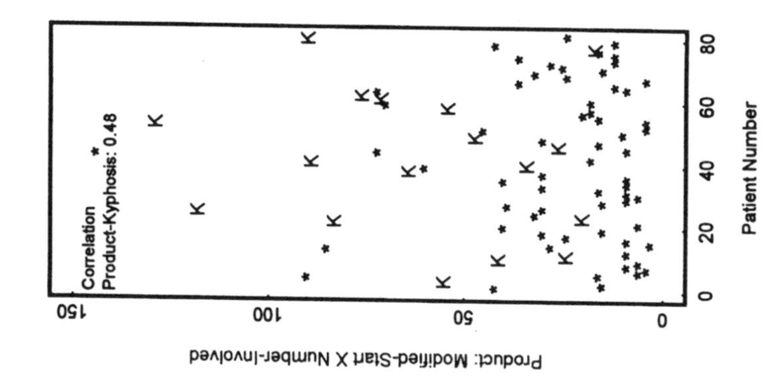

Figure 8.3: Scatter Diagrams with Transformed Variables

After making the transforms, we again ran the model many times using these new variables, Modified Age (MA), Modified Start (MS), the (Modified Start) × (Number) product (Prod), and the unmodified Number (N) variable. Runs were made with varying sample sizes and time assignments using a hazard model, and with changing sample sizes using a logistic model. The four variables were taken individually and in numerous combinations. It developed that the product term, for which we had high hopes, simply confused the issue. It generated fluctuations much larger than seen with the unmodified variables. It appears that there is no interplay between location and number of vertebrae involved. However, the remaining three variables, MA, MS, and N, produced excellent results. The relative risks remained just as stable as before and the P-values stabilized and all dropped well below an acceptable 5 percent level as the sample size approached the maximum number of eighty-one.

All of this is summarized in figure 8.4, where we show one typical run using the unmodified variables and a typical one with the modified factors. The three plots, from top to bottom, present results for the variables age, number of vertebrae involved, and the starting location. Each one has the results for both the original data (white markers) and the modified data (black markers). The major difference appears in the top plot for age and modified age. Before the modification, age (the white squares) was declared to be of no interest, the risk value is right at 1.0 for all sample sizes. Note though that the P-values (white triangles) are very poor, in the vicinity of 80 percent. Using the modification to make the relation between age and kyphosis linear, reveals that increasing (modified) age is a very positive factor in guarding against the problem. The black squares fall well below the null risk line and the P-values (black triangles) are consistent and move below the 5 percent level as the sample size grows.

You may note that the P-values are also less than 5 percent for the smallest samples. If this were a study in which time assignments or ordering in time were a real factor, that would be cause for some concern and show that "more work is needed." In this case, because the assignment of time slots is only an artifact used to explore the data, it is probably safe to ignore those occurrences.

Figure 8.4 presents a simplified condensation of the kinds of events that are associated with a good study. The results represented by the white markers are typical of the first tries at modeling or "fitting" the scatter diagram of test results. Some appear to be satisfac-

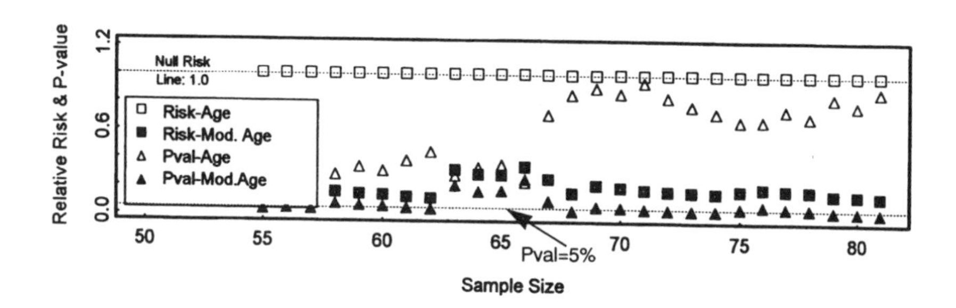

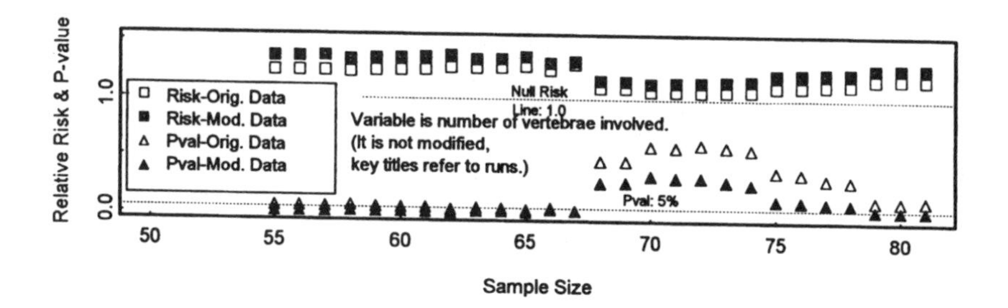

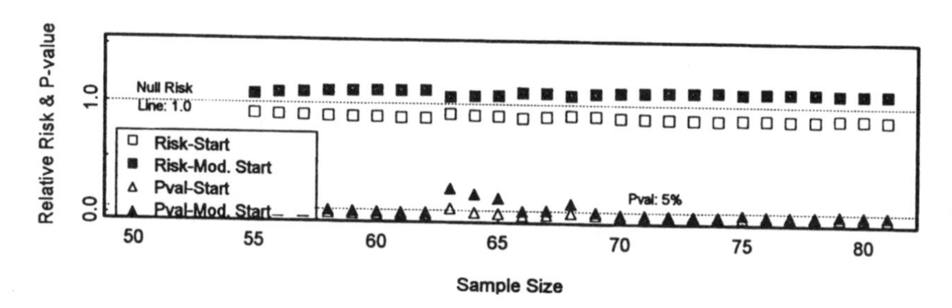

Figure 8.4: Typical Hazard-Type Result with Two Modified Variables

---

tory but others, the white triangles here, are not "well behaved," they jump around a lot and one is left with a feeling of uncertainty. On the other hand, the black markers are typical of results when "all the loose ends are tied down." They are all well behaved. The patterns of dots are smooth and they converge properly with sample size. These facts are very encouraging with regard to the integrity and value of the study.

In a real study effort, one crucial test remains. Recall that in lesson 3, once we had found formulas relating speed and such to deaths in car accidents, we used them to compare the model results with what actually happened, and then to make predictions as to what would happen under different circumstances. The same should be done here. Without going into the detail, suffice it to say that the results of the "good" runs discussed above did generate numbers in excellent agreement with the data. Of course, they should be used now to make predictions for the chances of kyphosis in future operations. The three critical variables, modified age, location, and number of vertebrae involved will be known in advance, so we can predict the chances of success for each case. Such an experiment should have been a planned part of the original study proposal, perhaps called phase B, to be undertaken only if phase A (the work we went through above) was successful.

So, working with an artificial hazard-type model and sample sizes was beneficial and we have identified two strong indicators of a good study:

- Small and consistent P-values for many variations of sample size and ordering.
- Consistent relative risks over a significant range of sample sizes as the maximum size is approached. A "settling" of the result.

Once again, figure 8.4 provides good examples of results which do (the black markers) and do not (the white markers) meet these criteria.

We must also add to this list the very important step of verification, including comparing results with the original data and performing a "phase B" experiment. The third criterion is then:

- Good agreement with experimental results.

This example of a good study is likely to be more of an ideal to be strived for than a model for conducting real research. First, it is extremely unlikely that any researcher will do the kinds of stability tests we used here. One mustn't mess with the sample, remember? Second, even if a researcher did do that, the public would probably not have easy access to the detailed results. We still wind up taking the researcher at his word, or worse yet, taking some columnist's word for the results of the study. Don't be too upset. In the next and last lesson, we will determine ways to evaluate the information we can get.

Note

1. John M. Chambers and Trevor J. Hastie, eds., *Statistical Models in S* (New York: Chapman & Hall, 1993), p. 200.

Lesson 9

So, Should I Drink the Coffee?

The best, quick answer to questions raised by published studies (Well, what about that coffee, should I drink it?), goes back over two millennia to Aristotle's golden mean, "moderation in all things." This is particularly true for new or single studies. As I have insisted, one study is worthless. (Keep in mind that the golden mean principle must also be applied to its application. I would not advise moderate use of Russian Roulette or of other things demonstrably dangerous.)

Unhappily, the alternative (not quick) reply to that question is not simple. In the last lesson we found that a fair answer is determined when study results are examined for the following qualities:

- Small P-values (or comparable measures of goodness) for many variations of sample size and ordering.
- Consistent relative risks over a significant range of sample sizes as the maximum size is approached. A "settling" of the result.
- Good agreement with experimental results.

Those first two mean simply that the researcher's tests on the quality of his own results must be consistent and repeatable. The third one is plain common sense, and means that predictions should be verified. Is drinking coffee really bad for most people? Make more tests and see, and be certain that all the studies agree with the first one or have good reasons for not agreeing.

But, we have to admit that no individual or even most groups can conduct this research themselves. That luxury is reserved for the people doing the studies. What each of us can do, however, is to make the information conveyed in studies usable by paying attention to their fallout.

Studies that do not exhibit the qualities above will inevitably lead to numerous contradictory reports. As proof, just call the long list of past "medical discoveries" that have faded into the dust. Recall the resounding claims for or against coffee, sugar, saccharin, oat bran, fish oil, salt, the carbohydrate diet, the Scarsdale diet, B-12 shots, alar, the Type A personality, or a martini before dinner, just to scan the surface of the heap.

There are important lessons to be learned from all these conflicting views. First and foremost, *don't panic,* and keep a close watch on your purse or wallet. If you are currently happy and healthy in your lifestyle, you can remain that way only by not jumping on all the bandwagons.

There is no more validity in the claims supporting the current popular diets than those linking eggs and cholesterol or power lines and cancer. There is as much conflicting evidence in the first group as in the second. If such issues are ever resolved, it will probably take years. As has been pointed out by the more calm and intelligent scientists and writers, the current battles against major diseases are much more difficult than those of the past. If that were not true, the current objects of research would have been replaced by others. The easy stuff was disposed of years ago. Until such issues are resolved, there is no more reason to go along with any recommendations than to ignore them. How can you judge whether the current advice will turn out to be another oat bran or fish oil fiasco?

All of the preceding can be condensed into the first rule for survival under the studies blitz:

1. Ignore all studies for at least several years, then get *interested* only if all the results have been the same.

Now recall my comments about the worth of a single study: It's practically useless. That is, making one study is about the same as throwing a pair of dice, once, observing that the result is say, ten, and proclaiming "When you throw dice, the number ten comes up." This bit of wisdom translates into the obvious corollary to the first rule:

2. Never read anything, or listen to anyone, if the opening statement resembles "According to one study. . . ."

In other words, treat new study reports as what they are, the result of a first-time inquiry into something—a first impression. If the topic interests you, watch for future developments and, unless they're consistent, you can probably safely ignore them. I don't have a count, but it is common knowledge that the majority of proposed theories in any field prove to be erroneous in the long run.

There is another source of public information that you may have seen but probably never really looked at. No new drug, prescription or otherwise, goes on the market without a thorough statistical study conducted under rules established by the Food and Drug Administration (FDA), designed to ensure product safety. Those studies are good, *now.* How many drugs have been introduced and then removed from the market since the rules governing approval procedures were revamped after the thalidomide fiasco of the 1960s?* The FDA and the pharmaceutical companies it watches appear to have some real statisticians. Their studies are summarized and included in prepackaged prescriptions and they are called Patient Information, or something similar. Take the time to read that little flyer when next you see one. You will be amazed at all the bad side effects that might arise from taking simple over-the-counter medication. Statements about the real chances of their occurring will be there also. Interestingly, those data sheets refrain from the glamorous terms such as "risk" or statements about increased chances. They are simple summaries of the raw results of the studies demanded by the government. It seems that the pharmaceutical people understand the fuzziness of fancy analyses and modeling, and have no desire to make the acceptance process even longer and more costly, so they just publish the facts; for example, 8,753 people participated in the trials. Of those, 710 experienced mild gastrointestinal discomfort, 14 developed headaches after taking the drug, and 3 became violently ill. There were no fatalities or complications recorded during the two-year follow-up period.

This simple, clean-cut presentation is all we need. We can read, understand, and make up our own minds. Ignore those columns with mountains of words comparing the apples and oranges of competing risk analyses. Which brings us to the next rule:

---

*Thalidomide was an approved treatment for morning sickness in pregnant women, but it turned out to be a cause of severe physical deformities in the fetus.

3. Distrust any claim that brags or moans about risk factors. Play the part of that stoic detective of early TV, Joe Friday, just look for the facts. Search for information similar to that contained in the Patient Information flyers. A lot of words about relative risks may be an excuse for ignorance.

Before winding down, there is a class of reports I have not touched on previously but which deserve some comment and lead to my last rule. These are the very serious, well-intentioned documents often called "bulletins" or "news releases" and are most readily available on-line to those with computer access and who want to spend the time. There are a fair number of such bulletins, but I will cite only two as being representative.

The Centers for Disease Control (CDC), maintains a number of these electronic bulletin boards, such as the *Monthly Vital Statistics* from the National Center for Health Statistics (NCHS), the *Morbidity and Mortality Weekly Report* (MMWR), and so on. These are generally factual summaries of current study reports, or even the reports themselves. This is one place to get some of the facts on current health issues. We can play Joe Friday to a degree here but any set of facts found will almost always be for only one study, so be careful. Keep in mind as well that this is raw data; it must be interpreted and may not be as straightforward as the pill-bottle insert we discussed above.

A second type of serious report is represented by the *Harvard Health Letter,* one of a few such publications maintained by the Harvard Medical School. These tend to be less factual or data oriented and more interpretive, in an effort to remove that burden from readers' shoulders. The presentations are calm, folksy, and designed to develop a warm, comfortable feeling. Those who are naturally critical of media health reports may be lulled into believing what these *Letters* recommend. Interpretations by definition are subjective, and the ones I've seen in this publication seem to follow the wave of the popular media but with some lag time and with broad summaries, which provide an air of reflection. This makes them more seductive, but the recommendations, still of the yo-yo type, are simply less frequent or violent in their changes of position on a topic. One, in 1995, acknowledging that cholesterol in eggs may not be bad after all, was almost apologetic. So, this last group of reports are not characterized by hype or hysteria but neither are they extremely helpful. In the final analysis, if a subject is of real concern, you must seek out the facts, e.g., the number of studies, the real numbers of people who died or were healed (out of how many total), and make your own judgment. Don't be upset or intimidated by others who disagree, some dangerous "high risk" items, such as automobiles, are usable and even necessary for many of us. Further, in these studies, it is often not clear that the item blamed has any risk. Watch for further tests; if they disagree, it is likely that the previously blamed item is not at fault.

So, unless you are really unhealthy, or just think that you are, you are your own best judge of how to treat yourself. Which brings us to the final rule:

4. Remember Aristotle and, if you're healthy and happy, keep doing what you do.

A Closing Wish

I hope you now have a greater understanding for those popular words, "Studies show . . ." and "More work will be needed. . . ." Whether you ever make any studies of your own or just continue to be shocked or amused by the latest reports, I hope this book has made you a bit more aware of just how vital it is to understand how and why studies are conducted.

And a final word to the student statistical-researcher. Before you ever begin any data collection, sit down and convince yourself, and then your worst antagonist, that you really know what you are doing. *After* you have done that successfully, you can begin your project.

Glossary

Association. As used in statistics, a generic term for any relation between two or more variables. It is more general than similar terms, such as correlation, because no specific type of relation is implied.

Case-control (study). A statistical study in which people are selected from two groups, one of afflicted subjects, the other of healthy subjects. Their past medical histories are examined and compared to look for causes of the disease under study. This is a backward-looking or retrospective study. See also **Cohort.**

Chi-square. One of many types of distributions. This is one that contains only positive numbers. See also **Distribution.**

Coefficient of determination. A measure of decrease in uncertainty resulting from an analysis. Formally, the reduction in the sum of squares (SS) of residuals (errors) resulting from some statistical analysis, usually a linear regression. It is computed as the ratio of the variance (uncertainty) that can be accounted for by a linear fit to the total uncertainty. The coefficient of determination is equal to the square of the correlation coefficient. See also **Correlation coefficient, Fit, SS,** and **Variance.**

Cohort (study). A statistical study in which people are selected from just one healthy group. The subjects are tracked as time progresses to follow and record their medical histories. Those in the group that develop a disease are then compared to those that don't to look for causes. This is a forward-looking or prospective study.

Confidence. As used in the text in "confidence for the confidence interval," for example, it is my name for the confidence coefficient. It may be thought of as the chance that the width of the confidence interval is "reasonable." See also **Confidence coefficient, Confidence limit,** and **Confidence interval.**

Confidence coefficient. When an estimate of a sample, say, the average, is made, a region around that estimate, called the confidence interval, may also be stated. If more than one sample is taken, then every sample will have its own confidence interval as well as its own estimate. The confidence coefficient is the probability that some percent (often 95) of all these additional confidence intervals will include, or cover, the first estimate of the average. See also **Confidence interval** and **Confidence limit.**

Confidence interval. A region within which the true value of some statistic, say, the average, is expected to fall. The confidence interval is often stated as being plus or minus so many percent of the estimate. So, if the estimated average is 100, but the confidence interval runs from 90 to 110, we could say the estimate is 100 plus or minus 10 percent.

Confidence limit(s). The number(s) defining the confidence interval. These are often expressed as a percentage of the estimate and then incorrectly referred to as the error in the measurement or the error in the survey. The measurement as reported is correct, it is the number calculated from the data and if the error were known, the number should have been corrected. What is meant is that the measurement is thought to be accurate only to within the stated percentage. Strictly, the confidence limits define a range that has meaning only in terms of results of other similar surveys. See also **Confidence coefficient** for the correct interpretation.

Confounder. A variable or factor that accounts for at least part of the effect being studied, but is often irrelevant to the study. Treatments that affect heart disease are confounded by age. That is, older people are more prone to heart disease, and so, in an evaluation of estrogen therapy and its effects on heart disease, for example, age could obscure other effects. Methods exist to account for confounders and remove their effects from the analysis, but only if they are included in the study and tested for interactions with the factors of interest. If not included, they may severely distort the study results.

Confusion index. My term for **significance level.**

Control group. A portion of the subjects in a study for which no special treatment is applied. This group is expected to behave or perform in the normal and customary manner, thus providing a reference with which to compare the performance of the group treated.

Correlation. A statistical relation between variables or factors of interest. Correlation is often mistakenly thought to imply a cause and effect relation, which it does not. See also **Correlation coefficient.**

Correlation coefficient. A number between –1 and +1 that describes the statistical relation between two factors of interest. It describes the portion of the total variability that is common to both of them.

Deviate. An older, less common name used for any factor of interest in a study, but often implying a specific value of the factor. See also **Variate** and **Variable.**

Distribution. A description of how things are naturally arranged on some scale. For example, the distribution of the height of people shows what fraction of people fall in each (small) group of heights. One group may include all people from 5 feet to 5 feet, 1 inch, for example.

Double blind. An experimental procedure in which neither the experimenter nor the subject knows which of a possible number of treatments a subject is receiving. This tends to reduce experimenter bias.

Expectation. Another name for average or mean.

Expected proportion. The average of a number of measured proportions (fractions).

Expected value. Another name for average or mean.

Fit. An informal way of referring to a line drawn through a set of points or scatter plot and used to describe the points in the diagram. The line is said to be "fitted" to the points. See also **Scatter plot.**

Gaussian. One of many types of distributions. Also known as normal. This is one of the types that may contain negative as well as positive values.

Hawthorne effect. That characteristic of people that makes them behave in an unusual or atypical fashion if they know or suspect that someone is watching.

Heterogeneous. Term describing a characteristic that is variable among members of the group.

Homogeneous. Term describing a characteristic that is very similar or even identical among members of the group.

Hypothesis. An assumed fact about something. Stated in a manner suitable for testing, as an item to be proved or disproved. A conjecture subject to verification. Often the stated hypothesis is made in a negative sense, e.g., "there is no difference" between the items being tested, and so may be referred to as the null hypothesis.

Hypothesis testing. The formal procedures for estimating the probability that the stated hypothesis is correct.

Intrinsic variation. An informal name for the variability or variance that is commonly found in some characteristic of a group. The colors of tree leaves in the fall exhibit a large variance and could be said to have a large intrinsic variation.

Linear analysis. Any form of analysis that makes the assumption that all variables are related to each other in simple ways, often just directly or indirectly proportional. Only the basic arithmetic (linear) operations are used to describe relationships; i.e., no squares, logarithms, or other "nonlinear" operators are permitted. Also called straight-line analysis.

Linear regression. Regression restricted to straight-line analyses. See also **Regression.**

Logit. Name used for the logarithm of a ratio of two odds.

Mean. The common average value of a group of numbers.

Multidimensional. Refers to studies that consider more than two variables.

Multivariate. See **multidimensional.**

Nonlinear regression. Regression using nonlinear mathematical equations. See also **Regression.**

Normal (distribution). Common name for a Gaussian distribution.

P-value. In hypothesis testing, the probability that the hypothesis is correct.

Pilot study. An experimental undertaking carried out prior to a full-scale test or survey. The pilot is used to assess the effectiveness of the survey.

Placebo. A harmless and useless dummy pill or treatment used in a part of a sample of subjects to make all the subjects think they are being treated equally. This procedure tends to wash out or nullify any Hawthorne effect.

Probability. A number between zero and one, expressing the relative frequency of occurrence of one event in a collection of events; e.g., the expected number, expressed as a fraction, of times a "seven" will come up if a pair of dice is thrown a great many times.

Proportion. A fractional part of something, often expressed as a percent.

Prospective. See **Cohort.**

Random variable. A variable that defies prediction of the next value; i.e., one cannot tell what the next number in a sequence will be. A variable that has no apparent pattern or order to it.

Regression. The process of finding mathematical equations, often for straight lines, that approximately describe the pattern or the behavior of one variable when others are changed. See also **Fit.**

Residual. The unexplained deviation(s) (errors), left in a set of data after all analyses are completed.

Retrospective. See **Case control.**

Sample. A collection of objects that have been identified to be measured or evaluated in a study. A sample should be representative of all the objects in the population from which it is drawn. In medical studies the sample is the group of people chosen as subjects in the study.

Sample survey. The process of measuring or evaluating the members of a sample, e.g., taking a poll.

Scatter plot. A graph constructed by using small markers (dots) to show that a data point exists at those coordinates. In general, the outline of the image(s) formed by the dots is not a smooth line. The points are scattered over the plane.

Sensitivity analysis. An examination of the effects of small changes in parameters used to estimate the robustness of a design or test result.

Sigma. A name applied to the standard deviation of the mathematically useful concept of an ideal distribution; e.g., a perfect normal distribution. Note that no real-life distributions exactly match the ideal mathematical descriptions used. Standard deviation, an identical concept, is meant to refer to real-life distributions. See also **Standard deviation.**

Significance level. When a study concludes that a relation has been found between two things, e.g., drinking and heart disease, the significance level is the chance that the conclusion is wrong. Formally, in hypothesis testing, significance level is the probability that the null hypothesis is correct. (In the usual situation, the null hypothesis states that there is no difference between the two things being examined.) Conversely, the significance level is also the probability that a claim that there is a difference is wrong.

Sum of Squares (SS). The sum of the squares of all deviations from some reference number. Formally, the differences between the actual values and the mean value, squared to make all the numbers positive, and then summed. The SS is a measure of the variability in a set of numbers and is used in many statistical calculations such as in computing variance, correlation coefficient, and so forth.

Standard deviation. The average value of the variations of numbers around their average. It measures the average difference from the average. Standard deviation is the square root of the variance. See also **Variance.**

Straight-line analysis. See **Linear analysis.**

Survivor function. A plot showing the percentage of subjects in a study who are still healthy at any point in time during the study period.

Uniform. One of many types of distributions. Numbers from a uniform distribution all have the same (uniform) chance of being selected. The distribution is "flat" over its whole range.

Variable. A name used to denote any factor of interest in a study. Variables may be numbers, such as people's heights, or qualities, such as the color of their eyes.

Variance. The average of the sum of squares (SS). A measure of the squared deviations or variations from the average value. Standard deviation is the square root of the variance.

Variant. A name used to denote any factor of interest in a study, but often implying a specific value of the factor. One variant of height of people might be 65 inches. One variant of eye color might be brown.

Variate. A synonym for random variable.

Appendix

An Annotated Bibliography

The following references are grouped in four categories to facilitate the reader's choices.

The first category lists a few of the numerous introductory texts suitable for a beginning undergraduate or an honors high school program. This includes at least one volume (on straight line analysis) that is relatively advanced but so lucid it belongs here. Books akin to *Statistics for Dummies* are not considered. The present volume fills that role.

The second category cites intermediate level material, designed to polish the skills and hone the understanding well beyond an introduction to the subject.

The third group comprises fundamental source material: journal papers and specialized texts. For the most part these represent rather heavy work. The semiskilled should avoid them, yet they are often the only sources reliably defining the subject.

Category four is miscellaneous. It includes references cited in the lessons and an odd collection of particularly interesting study reports, editorials, and software handbooks.

Category One

Acton, Forman S. *Analysis of Straight Line Data.* New York: John Wiley & Sons, 1959.

Chatfield, C. *Statistics in Technology*. London: Chapman & Hall, 1970.
Drake, A. W. *Fundamentals of Applied Probability Theory*. New York: McGraw-Hill, 1967.
Fraser, D. A. S. *Statistics: An Introduction*. New York: John Wiley & Sons, 1958.
Wadsworth, George P., and Joseph G. Bryan. *Introduction to Probability and Random Variables*. New York: McGraw-Hill, 1960.

Category Two

Box, George, E. P., William G. Hunter, and J. Stuart Hunter. *Statistics for Experimenters*. New York: John Wiley & Sons, 1978.
Cochran, William G. *Sampling Techniques*, 2d ed. New York: John Wiley & Sons, 1963.
Feller, William. *An Introduction to Probability and Statistics*, Vol. I, 2d ed. New York: John Wiley & Sons, 1950.
———. *An Introduction to Probability and Statistics*, Vol. II. New York: John Wiley & Sons, 1966.
Hald, A. *Statistical Theory with Engineering Applications*. New York: John Wiley & Sons, 1952.
Hinkelmann, Klaus, and Oscar Kempthorne. *Design and Analysis of Experiments*. New York: John Wiley & Sons, 1994.
Schlesselman, James J. *Case-Control Studies Design, Conduct, Analysis*. New York: Oxford University Press, 1982.
Trivedi, Kishor, S. *Probability and Statistics with Reliability, Queuing, and Computer Science Applications*. Englewood Cliffs, N.J.: Prentice-Hall, 1982.

Category Three

Brown, Byron W., Myles Hollander, and Ramesh M. Korwar, "Nonparametric Tests for Independence for Censored Data, with Applications to Heart Transplant Studies" in *Reliability and Biometry: Statistical Analysis of Lifelength*, Frank Proschan and R. J. Serfling, eds. (Philadelphia: Society for Industrial and Applied Mathematics, 1974), pp. 327–54.
Cox, D. R. *Analysis of Binary Data*. London: Chapman & Hall, 1970.
Crowley, John. "Asymptotic Normality of a New Nonparametric Statistic for Use in Organ Transplant Studies." *Journal of the American Statistical Association* 69 (December 1974): 1006–11.
Crowley, John, and Marie Hu. "Covariance Analysis of Heart Transplant Data." *Journal of the American Statistical Association* 72, no. 357 (March 1977): 27–36.
Gail, Mitchell. "A Review and Critique of Some Models Used in Competing Risk Analysis." *Biometrics* 31 (March 1975): 209–22.

Greenland, Sander. "Quantitative Methods in the Review of Epidemiologic Literature." *Epidemiologic Review* 9 (1987): 1–30.

Kalbfleisch, John D., and Ross L. Prentice. *The Statistical Analysis of Failure Time Data.* New York: John Wiley & Sons, 1980.

Terry, M., et al. "Martingale-Based Residuals for Survival Models." *Biometrika* 77, no. 1 (1990): 147–60.

Category Four

Bailer, John C., III. "When Research Results are in Conflict." *New England Journal of Medicine* 313, no. 17 (October 1985): 1080–81.

Chambers, John M., and Trevor J. Hastie, eds. *Statistical Models in S.* London: Chapman & Hall, 1993.

Clark, David A., et al. "Cardiac Transplantation in Man, VI. Prognosis of Patients Selected for Cardiac Transplantation." *Annals of Internal Medicine* 75, no. 1 (July 1971): 15–21.

Fitzgerald, Karen, et al. "60 Hertz and the Human Body, Parts 1, 2, 3." *IEEE Spectrum* 27, no. 8 (August 1990): 23–35.

Gail, Mitchell H. "Does Cardiac Transplantation Prolong Life?" *Annals of Internal Medicine* 76, no. 5 (May 1972): 815–17.

Gordon, Tavia. "Further Mortality Experience among Japanese Americans." *Public Health Reports* 82, no. 11 (November 1967): 973–84.

Halpern, Diane F., and Stanley Coren. "Do Right-Handers Live Longer?" *Nature* 333 (1988): 213.

———. "Left-Handedness: A Marker for Decreased Survival Fitness." *Psychological Bulletin* 109, no. 1 (1991): 90–106.

Hulley, Stephen B., et al. "Health Policy on Blood Cholesterol, Time to Change Directions." *Circulation* 86, no. 3 (September 1992): 1026–29.

Keys, Ancel. "Epidemiologic Aspects of Coronary Artery Disease." *Journal of Chronic Diseases* 6, no. 5 (November 1957): 552–59.

Leeper, Edward, and Nancy Wertheimer. "Electrical Wiring Configurations and Childhood Cancer." *American Journal of Epidemiology* 109 (1979): 273–84.

Levine, Robert V. "The Pace of Life." *American Scientist* 78, no. 5 (September–October 1990): 450–59.

Perry, Tekla S. "Today's View of Magnetic Fields." *IEEE Spectrum* 31, no. 12 (December 1994): 14–23.

Pike, M. C. "A Method of Analysis of Certain Class of Experiments in Carcinogenesis." *Biometrics* no. 22 (1966): 142–61.

Statlib. *On-Line Library for S-Plus Software Users.* Maintained by Carnegie Mellon University: <statlib@lib.stat.cmu.edu>.

Springer, Ilene. "Castelli Speaks from the Heart." *AARP Bulletin* 33, no. 5. Washington, D.C.: American Association of Retired Persons, 1992.

Stampfer, Meir J., et al. "Prospective Study of Postmenopausal Estrogen Therapy and Coronary Heart Disease." *New England Journal of Medicine* 313, no. 17 (October 1985): 1044–49.

Turnbull, Bruce W., Byron W. Brown Jr., and Marie Hu. "Survivorship Analysis of Heart Transplant Data." *Journal of the American Statistical Association* 69 (March 1974): 74–80.

U.S. Department of Commerce, Bureau of the Census. *Statistical Abstract of the United States.* Washington, D.C.: GPO, 1920–1995.

Wilson, Peter W. F., Robert J. Garrison, and William P. Castelli. "Postmenopausal Estrogen Use, Cigarette Smoking, and Cardiovascular Morbidity in Women over 50." *New England Journal of Medicine* 313, no. 17 (October 1985): 1038–43.